Why the Dose Matters

Each day we are exposed to a myriad of natural and human-made chemicals in our food, drinking water, air, soil, at home or at the workplace—pesticide residues, food additives, drugs, household products—but how can we gauge the human health risk posed by these chemicals? Should we believe the somber headlines that depict a serious threat to humans and the environment, or should we follow the reassuring voices of others who claim that the angst is totally unfounded?

Why the Dose Matters: Assessing the Health Risk of Exposure to Toxicants uses a rational, science-based approach to explain in plain language that a quantitative view is key for understanding and predicting potentially toxic effects of chemicals.

Key Features:

- Explains the basics of toxicology in easily understandable terms.
- Includes numerous examples.
- Clears up common misconceptions and dispels myths.
- Provides take-home messages for each chapter.

This book is aimed at interested laypeople. It uses numerous examples to illustrate the basic concepts and ensure that the reader will get a better understanding of why not only the hazard but also the overall exposure will determine whether some chemicals pose a serious risk while others are of little or negligible concern.

Why the Dose Matters
Assessing the Health Risk
of Exposure to Toxicants

Urs A. Boelsterli

CRC Press
Taylor & Francis Group
Boca Raton London New York

CRC Press is an imprint of the
Taylor & Francis Group, an **informa** business

Designed cover image: Shutterstock | Danusya.

First edition published 2023
by CRC Press
6000 Broken Sound Parkway NW, Suite 300, Boca Raton, FL 33487-2742

and by CRC Press
4 Park Square, Milton Park, Abingdon, Oxon, OX14 4RN

CRC Press is an imprint of Taylor & Francis Group, LLC

ISBN: 978-1-032-38765-9 (hbk)
ISBN: 978-1-032-38764-2 (pbk)
ISBN: 978-1-003-34666-1 (ebk)

DOI: 10.1201/9781003346661

Typeset in Caslon
by KnowledgeWorks Global Ltd.

To the late Gerhard Zbinden, who taught me
the difference between hazard and risk.

To Carmen, who taught me that life is full of hazards and risks,
but that managing the risk is better than worrying about it.

Disclaimer

It is important to note that the readers should not base any personal decision about dosage of chemicals including pharmaceuticals, choice of consumer products, handling of household chemicals, or exposing themselves to foreign chemicals solely on the numbers given in this book; instead, the latest original sources should be consulted, for two major reasons. First, as science advances and new knowledge is being gained, dosage or exposure limits can change, and second, even though the author has put maximum effort in providing accuracy, errors are not inevitable.

Contents

PART IV THE FUTURE

List of Figures and Tables

Preface

In 2021, the Swiss people had to vote on whether to accept or dismiss a referendum initiated by a group of environment-conscious citizens. The initiative's theme was (I'm paraphrasing): should there be a total ban on the use of all synthetic pesticides across the country? The heated debates on TV, in the media, at the cracker-barrel, and across garden fences are long over, and I really do not want to reopen the case, nor open a new can of worms (by the way, the initiative was rejected). What made me ransack my brains over, though, even long after that political decision was made, were two nagging questions:

First, why do a lot of people believe that everything synthetic is bad and everything natural is good? And second, why do a lot of people believe that a chemical must be toxic just because it is there?

Before the vote, the country had never seen so many self-appointed toxicology pundits, suddenly mushrooming from nowhere. But what was more astonishing was the fact that most of the acrimonious arguments (from both advocates and opponents) were often fueled by emotions, bold assumptions, beliefs, and knowledge from hearsay rather than being based on rational facts and sound knowledge. If you let yourself become entangled in a discussion, you were either reviled as an idealistic green activist or stigmatized as a lobbyist for the pesticide industry. Where was the science? Fallen by the wayside?

The science of toxicology is dauntingly complex. Not only because there is a lot of chemistry to grapple with, but also because it involves an understanding of the underlying biology and pathophysiology, pharmacology, environmental sciences, and an insight into risk assessment and risk management. Toxicology clearly is multi-disciplinary, and it must therefore be quite confusing for a non-scientist to keep track of everything and assemble it to the big picture. You can't blame the laypeople when a discussion gets emotionally charged while the scientific, rational arguments are few and far between.

Part of the blame probably must go to the scientists themselves (I'm one of them)—why do they hide in the ivory tower, behind their hard-to-understand publications and scientific meetings, instead of getting the message across to the public in easily understandable language while not distorting the facts? Believe me, it's not the arrogance of the adepts. It's mostly a lack of time, and often the frustration that things have gotten too complicated to be explained in simple terms. (Which isn't easy; Einstein once said, "Make everything as simple as possible, but not simpler.")

It was on one of those days that the seed for this book project was planted in my mind.

This book is targeted at interested people who have no special training in the field but who would like to see through the mist of this complex area without having to whack through the thick jungle of scientific jargon (and I promise, you won't find any chemical equations or molecular structures). It is not a comprehensive reference book (there are many outstanding textbooks on toxicology on the market)—instead, it illuminates just a couple of key aspects that I think will help better understand the complex interactions between chemicals and humans.

Finally, I hope this book can convey some of the fascinations that's inherent in studying science, and I would like to share with the reader some of the passion I feel while further exploring the field of toxicology.

About the Author

Urs A. Boelsterli, Ph.D., FAASLD, is an Emeritus Professor of Toxicology at the University of Connecticut School of Pharmacy, where he also held the first Boehringer-Ingelheim Endowed Chair in Mechanistic Toxicology. Prior to this, he was head of the Toxicology Program at the National University of Singapore and ran a research lab at Roche, Basel, Switzerland, and the ETH in Zurich. He is retired and currently lives in Switzerland with his family.

1

INTRODUCTION

Have you ever watched the news or read a newspaper article and wondered why we are still alive?

Sinister reports about a new (or a notorious) bad chemical found in the food we love most, or a poisonous compound detected in the air or water, have been galvanizing us for years. Here we go again: another pesticide lurking on the surface of your breakfast berries. A nasty, cancer-causing chemical mysteriously showing up in the lake where you regularly take a swim. Blood-clotting nanoparticles in your white toothpaste. Or, worse even, the pills you have been swallowing for years turn out to be highly dangerous for your health, killing people. Seriously?

You are now immediately convinced the faint headache you've been experiencing lately must be a harbinger of a stroke, due to the painkiller you ingested last week. And what you thought was merely slight indigestion probably is a liver tumor caused by an apple that you forgot to peel. You panic. The world is crazy—after all these ominous reports, what the heck can you eat nowadays without getting sick? Is there still anything at all to be considered safe? Eventually, you decide to skip meals, refrain from breathing the ambient air, and not touch anything on this planet. Problem is, by averting being exposed to all these allegedly bad chemicals, you won't survive. Guaranteed.

If this sounds sardonic, it is not my intention. As a toxicologist, I'm certainly not trivializing the potential danger of certain chemicals—quite the opposite. All I want to do with this book is to make the reader aware of the difference between a real risk or potential danger imposed by toxicants and those irrational, apocalyptic scenarios which more than often are nothing else than sensationalism, partly due to a poor understanding of the principles of how a chemical interacts with our body. Make no mistake, there are highly toxic chemicals in our environment, food, air, and water, which can potentially be harmful, even fatal. It is the toxicologists' task to call attention to those risks. The good news, however, is that not everything that bears the hastily added label "toxic" will invariably be dangerous to

DOI: 10.1201/9781003346661-1

our health. Toxicity is a relative term, and the key question is: how much of it is too much?

If somebody asked me what I consider the single most important take-home message for the reader who is patient enough to read their way through this entire book, I would answer without hesitation that it is the awareness that the science of toxicology is a *quantitative* discipline.

Let's go back a few centuries when a doctor of the Renaissance who called himself *Paracelsus* came up with a simple, but brilliant idea that in most parts is still valid today: it is the dose that makes the poison. More accurately, the net *exposure* to a chemical determines its toxicity—much more on this (keep going).

However, we must take a closer look at this concept. It is not so simple that we can just say, the more of something the worse for us. Any chemical (an environmental pollutant, a drug, something in our food) evokes a certain reaction in our body, a response which, in some cases, may wind up to be bad for us. However, it's not only the chemical that does something to our body—the body also does something (a lot, actually) to a chemical—in many cases, our body can transform it to a different compound that can be less or, in some cases, even more harmful. A part of this book discusses that in more detail.

Each day, we are being exposed to a zillion of different chemical compounds, and new ones are being made every year. For many of these synthetic chemicals, we have insufficient information as to their potential human health risk—because we simply don't have the resources (money, experts in the field, and time) to thoroughly investigate every single chemical on this planet; we must set priorities. So, have we been digging our own graves? Are we falling victim to our own technical creations, like the sorcerer's apprentice? Are we being flooded by too many synthetic chemicals so that we've lost track of their potential risks?

Let's wait for a second—unlike the popular misconception that the "bad" chemicals must be human-made, while everything natural, by definition, must be "good," the real picture is different. Some of the most potent toxic chemicals occur in Nature. Think of the many different, highly toxic compounds in plants (that, during evolution, provided a big survival bonus from being eaten by animals).

Think of toxic metals in the soil; the dangerous fumes arising from forest fires; the radioactive, naturally occurring gas, radon, which can contribute to lung cancer. Even oxygen, essential for all aerobic life, can be toxic!

Fortunately, we are not completely and utterly at the mercy of potentially toxic chemicals in our environment, food, or air. Animals and humans could not have evolved and survived to the present day if we were not sufficiently protected by several layers of biological firewalls and molecular defensive shields. We are tough and can handle quite a bit.

And let's not forget that we've come a long way. Toxicologists worldwide have the mission to study and analyze the effects of chemicals on living organisms and protect our citizens and the environment from potentially harmful effects with ever-improving technologies and testing. There were terrible disasters in the past that hopefully will never happen again; examples include the tragedy with thalidomide, a pill against morning sickness, that lead to malformations in numerous newborn children in the 1950s, or the generous use of the insecticide, DDT, during World War II, posing a threat to wildlife and birds for decades. More recent examples are perhaps less dramatic but still have had a tremendous impact on human health. Why could all that happen in the first place? There has been a lot of finger-pointing at the alleged ineffectiveness of the scientists and their poor methods of predicting the toxicity of "bad" chemicals. But in retrospect, we always know better, don't we? The truth is, it's the other way around: because at that time we did not yet have the tools we have nowadays (including the rapidly growing panel of sophisticated testing methods for toxicity, or the extremely sensitive methods for detecting even traces of chemicals) such "accidents" could happen. We are much better off today than even a few years back—but we still need to be vigilant about potential new threats.

We need to look at the toxicity of chemicals and the potential risk they might pose with a rational mind and not lose sleep over problems that are artificial and based on emotions, amplified rumors, fake news, or assumptions. We too often seem to forget that a chemical does not pose a serious human health risk *just because it's there*.

Now let's get started and have a closer look at some of these concepts.

TAKE-HOME MESSAGE

- We are being exposed to thousands of different chemicals each day—some of them highly hazardous—but everything doesn't harm you just because it happens to be there.
- The science of toxicology, describing the effects of chemicals on humans, animals, and the environment, is a *quantitative* discipline.
- *For that matter*: Don't let the rumors and flashy headlines upset you—get down to basics. The better you know the facts, the easier can you gauge the risk.

PART I
THE CONCEPTS
Hazard and Exposure

2

WHAT DOES "TOXIC" MEAN?

Pegging a Chemical *a Priori* as Bad?

Humankind would not have survived as a species if we were exposed naked and unprotected to the plethora of harmful chemicals abundant in our environment. They're everywhere. I'm not talking about the extremely powerful toxins of certain microorganisms, plants, or animals—it is obvious that, if exposed to them, minute amounts of them can be fatal—I'm talking about all those less flashy chemical compounds in our foods, in the air, the soil, water, that, depending on the unfavorable circumstances, could harm us in one or several ways. Some of them can pose a real risk, others pose a perceived threat that, if scrutinized, is not realistic under normal, everyday conditions.

So, what does "toxic" mean? Classical dictionaries define the term as something poisonous (like a toxic mushroom), very harmful (such as a toxic exhaust fume), or really bad (like toxic relationships, but I'm not going there; I'll confine myself to toxic chemicals). Something worse than just noxious—something that can cause mild or severe harm, and everything in between.

It is often said that *everything* can be toxic, depending on how you define the term—which might be true for someone who has an extravagant imagination. As mentioned above, even oxygen under certain conditions can damage our cells, but nobody would consider molecular oxygen a "toxic gas" (unless you are in a pressurized chamber, recovering from carbon monoxide poisoning or a decompression accident after scuba diving, and you receive pure oxygen for a duration that's too long). Similarly, certain vitamins, although by definition vital for us, can cause harm, and yet hardly anybody thinks about kidney stones when they pop a multivitamin pill. At the other end of the scale are highly dangerous chemicals that can be lethal in a short time. Nobody would argue against the fact that a cobra toxin is toxic. So, is the term "toxic" woolly and vague, too nebulous to be used to characterize a chemical? If so, how about "very toxic" or "not toxic"? (Figure 2.1.)

But it's more than just semantics. Not only is "toxic" a relative term but it also needs the addition of a few words that describe the

DOI: 10.1201/9781003346661-3

"Very toxic?" ... "Quite safe?" ... "Sort of dangerous?" ...
" Relatively harmless?"... " Pretty bad for you?" ...
" Innocuous?" ... " Kind of harmful?"

Figure 2.1   Safety and toxicity are relative terms.

conditions and special circumstances under which a person might be exposed to that toxic stuff. Most importantly, it is the exposure that makes a chemical to be toxic or not (you may have guessed that already). We will talk about the important quantitative aspect of toxicity throughout the book.

So, to say: "An aspirin is toxic! Keep your hands off!" is nonsense.

But to say: "Prolonged intake of high doses of aspirin, e.g., one gram (g) each day, may lead to gastric irritation and peptic ulcers, while daily intake of low doses such as 80 milligram (mg) may be beneficial in patients with a history of coronary heart disease, and thus outweighs the risk" while sounding wordy, academic, even pompous, would be more accurate. Although you risk that your dialog partner unexpectedly wants to change the subject.

Of Industrial Toxicants, Plant Toxins, Magic Poisons, and Snake Venoms

For starters, and since we are talking about the choice of the correct words and the emotions they evoke, here are a few examples.

Did you ever notice the subtle difference in tone and implication according to who is talking about the same thing? When the research toxicologist talks about *toxicity evaluation* of a product, the manufacturer calls it *safety assessment*. When a chemical company talks about *agents for crop protection*, the environmental scientist calls them *pesticides* (the mere sound of the word gives you the creeps). When the package insert lists all the rare to extremely rare adverse effects of a pharmaceutical drug, for better transparency and liability reasons, the patient is scared witless, never touching the pill bottle again and rather enduring the horrible pain from the disease. It's all about someone's angle of view. It's the unspoken meaning that percolates (do you know of any country whose government has a department of aggression? All have a department of defense.)

When talking about the toxicity or safety of chemicals in a rational way, we need to get rid of the emotional component and keep to the facts, without sugarcoating an existent issue, but also without demonizing something less perilous.

It's time for a few definitions, to complete the confusion.

A *toxicant* is a human-made, synthetic chemical that's potentially harmful, or a product arising from human activity (industry, traffic, burning fossil fuels, etc.). In contrast, a *toxin* is a naturally occurring potentially harmful substance produced by animals, plants, bacteria, or fungi. Finally, if an animal is able to actively deliver the toxin it produces (e.g., a snake bite or a wasp sting), the toxic substance is termed as *venom*.

How about *poisons*?

The term is a bit outdated, except for its use in mystery novels, spy thrillers, and historical anecdotes. It usually refers to a chemical that causes severe injury or death, in relatively small amounts administered. This may occur either accidentally (e.g., a poisonous mushroom) or intentionally (e.g., a poisonous arrow). Nowadays, the word is used in colloquial speech rather than in scientific literature. In a gripping suspense novel, the amateur sleuth discovering the "toxicant" in the victim's drink would sound equally bizarre as the description of a "poisonous" action of an anti-anxiety drug in a scientific paper.

Again, note the mood-inducing undertone in these sentences: "Pesticides are poisoning our drinking water," says the alarmed environmentalist, versus "The levels of glyphosate in our drinking water are way below the safe daily intake limit," says the hobby gardener.

So, if toxicity is a relative term as we've just seen, defined as the *potential* of a chemical to cause adverse effects in animals and humans, what does ultimately determine the real toxicity of a chemical? On a more practical level: how "toxic" for the consumer is, say, a new pesticide that has just been approved by the regulatory authorities?

Figure 2.2   The major determinants of the toxic response.

The question may sound finicky, but the answer is simple. Two major factors determine the toxic response (there are other factors, stay tuned): on the one hand, it is the inherent "power" of a chemical to cause harm (termed *potency*), and on the other hand, it is the *exposure* of an organism to that chemical (Figure 2.2). For example, a bacterial toxin may be extremely dangerous because a tiny amount of it might kill us (high potency), but if we never encounter it (no exposure) it wouldn't harm us at all. Alternatively, a chemical compound with a low, negligible potency may be safe for the average consumer although we consume it each day.

Therefore—*potency times exposure* determines the outcome (the toxic response).

Let's have a closer look at these two measured variables.

Potency

If you've ever tried to take a hearty bite of an innocuous-looking chili pepper, you will immediately understand the concept of potency (and you can skip the rest of this chapter). It's trivial knowledge that different chilis have different inherent power of lighting a hell fire in your mouth. They range from the juicy, harmless bell peppers in your salad to the slightly spicy pepperoncini on your pizza, on to the fiery jalapeño peppers in your salsa, and—if you have a masochistic proclivity and take the next day off from work—all the way to the extremely hot and treacherous habanero chilis. There is even a quantitative, though subjective, measure of the pungency in the different kinds of chilis: the infamous Scoville scale named after an American pharmacist in the early 20th century (nowadays, the pungent principle is quantified by modern, highly sensitive analytical techniques). The higher the number of

the Scoville Heat Units, the hotter the chili. So, if you are comparing different chilis with each other, you can rank them according
to their potency. Since the exposure to the different chili peppers
is identical, assuming you take an equal bite size for all, it is the
potency that determines the degree of that burning sensation on
your tongue (the response).

This red-hot metaphor may be plausible, but there is a catch. The
chemical that causes the tingling sensation or explosion, in your
mouth, is *capsaicin* (probably an evolutionary advantage for the pepper plant as it would deter animals from eating the plant). But while
we can rank the different types of chili peppers according to their
potency, the underlying cause is always the same. What varies is the
amount of capsaicin in the different chilis. Now imagine you compare
different chemicals according to their potency to induce, e.g., cell
damage, injury to an organ, or death, or whatever might be the endpoint you take as a reference—and now you are able to rank different
chemicals with respect to their potency. And again, this can span a
wide range.

To illustrate this, let's look at a short list of chemicals in Table 2.1.
The endpoint in this case is lethality (the ability of these compounds
to induce death in laboratory mice, given as a single dose).

The metric system is used in science. For units and abbreviations for
weight, volume, or concentrations, see *Appendix 1*.

The table demonstrates that the potency of botulinum toxin,
one of the most potent toxins known (better known under its trade
name, Botox, which is used for cosmetic purposes) is 400 million
times greater than that of table salt with regard to that single endpoint,

Table 2.1 Potency of Different Chemicals to Produce a Toxic Effect (Lethality)

CHEMICAL	DOSE TO INDUCE DEATH*
Table salt (sodium chloride)	4,000
Morphine	900
Nicotine	1
Tetrodotoxin (puffer fish toxin, fugu)	0.1
Botulinum toxin (a bacterial toxin)	0.00001

*The numbers are the dose administered (in mg per kilogram body weight) that
would kill half of the exposed laboratory mice; this dose is called LD_{50}.
Note: These numbers are approximate figures. We will talk more about the
sense and nonsense of using the LD_{50} values later in this book.

lethality. Potency compares two or more compounds with each other; it's a relative term.

But here is the sixty-four-thousand-dollar question: what would you say when asked: what is the potentially greater human health risk for the average citizen, to die of the botulinum nerve toxin after food poisoning, or develop hypertension that is promoted by consuming daily too much table salt? I leave the answer to you—but you have recognized that a third factor comes into play—probability. The probability of being exposed to a chemical under average, realistic conditions in our daily life. (But I don't want to get ahead of myself—more of that is discussed in Chapter 17.)

Is It Safe? On Hazard and Risk

Is it safe to ride a motorbike?

No, says the paramedic who's staring at the fatal crash scene.

According to the US National Highway Safety Administration, more than 5,000 people died in a motorcycle accident in one year (2019).

Yes, says the Harley fan, if you are careful, share the road with the others, wear a helmet (and ride sober).

Both are right, of course, but they're talking about different things. The paramedic is talking about the *hazard*, and the biker about the *risk*.

Here is the thing: while all of us are aware of the difference between the two in the above example, people often seem to be blind about the distinction when it comes to the toxicity of a chemical. (Fortunately, the English language has two words for it; in German, e.g., both hazard and risk are simply called *Risiko*.)

So, if someone claims that a certain pesticide is not safe for humans, they can be both right and wrong. For example, a hypothetical insect-killing compound may pose a high hazard due to its neurotoxicity (damage to parts of the nervous system), but the risk may be low because humans would hardly ever come into contact with it, because the insecticide would be long-washed off at the time of harvest, or becomes diluted in the soil, or is rapidly degraded in the environment, and because the minute amount still residing on the veggie or fruit in the consumer's plate is too small to cause any adverse effect.

To summarize, hazard is defined as the *inherent ability*, or the *potential*, of a chemical to cause harm (think of the "FLAMMABLE" symbol on a gasoline truck). Risk, on the other hand, is defined as the *probability* of an adverse effect to happen based on both exposure and potency of the hazardous chemical under given circumstances (think of the "NO SMOKING" sign posted at the gas station).

Therefore, when a new drug, a food additive, or a household chemical, is labeled as "safe" for use, it usually has gone through the complex process of *hazard identification*, followed by a rigorous *risk assessment*—two terms that you will encounter frequently throughout this book.

Key to all these considerations about safety and risk, apart from the hazard, is the *dose* of a chemical—a term so pivotal that I've chosen to use it in the title of this book. This segues into the next chapter.

TAKE-HOME MESSAGE

- The word *toxic* is vague and often fraught with emotions; it is a relative term.
- The toxicity of a chemical is determined by its *potency* to induce a certain toxic response multiplied by the *exposure* to that chemical.
- Toxicology is the science that explores the adverse effects of chemicals on living organisms through *hazard identification* and *risk assessment*.
- *Bonus*: Never bite into a chili pepper without a glass of milk at hand.

3

PARACELSUS RELOADED

The Dose Concept

From Master of Alchemy to Founder of Toxicology

Should you ever be so lucky as to be able to visit the city of Basel in northern Switzerland, plan an extra day or two. The old town with its quaint houses, towering over the river Rhine, is charming, the art museums worth visiting, and a jaunt to the adjacent Alsace region in France will be unforgettable, not just for food aficionados. The real gem, however (at least for folks like you, I guess, since you have been reading so far into the book), is the *Pharmazie-Historisches Museum*. Why? you ask. Fusty rooms in an old building, stuffed with dusty jars, ceramics, mortars, and dried herbs?

Way off the mark!

You will immediately be drawn into the mystic world of alchemy, staring incredulously at the paraphernalia of a centuries-old laboratory— but the best part of it is that you will feel the spirit of *Paracelsus* (Figure 3.1). In the early 1500s, Paracelsus was a frequent visitor of that house in a steep and narrow alleyway in the old town of Basel. It was there that he met other big shots of the Renaissance period—an ancient version of a modern Center of Scientific Competence!

If you've never heard about this guy, it's time for a quick digression.

Aureolus Philippus Theophrastus Bombastus von Hohenheim, who later called himself Paracelsus, was born in Switzerland in 1493. Early in his life, he developed a keen interest in chemistry, pharmacy, and related sciences. He studied medicine in Italy and elsewhere and got his doctorate in 1516 at the University of Ferrara, Italy. But unlike other scholars during the Renaissance period who concentrated on ancient texts, Paracelsus followed a new school of thought, focusing on a holistic relationship of humans with Nature, God, and the cosmos. He traveled widely across Europe, even to Egypt, to further expand his knowledge, teach, and practice medicine. In addition, he developed several new principles in alchemy and was soon considered a scholar in astrology, theology, chemistry,

DOI: 10.1201/9781003346661-4

Figure 3.1 Woodcut from Paracelsus, *Astronomica et estrologia opuscula*, Cologne, 1567.

philosophy, and pharmacy (only a Renaissance person could be such a universal genius).

In 1526, Paracelsus was offered the position of town physician of Basel. He was also a professor of medicine at the University of Basel, where he lectured mostly in German instead of Latin so that less educated people could get his message. But according to historical tradition, he wasn't always the most easy-going fellow to get along with, openly snubbing his colleagues. He was a free spirit, a reformer, harshly criticizing conventional medicine, ridiculing academics whose titles seemed more important to them than knowledge and experience, and challenging anybody who refused to accept his new approaches. Sadly, he became bitter and frustrated, and his anger (as well as the ire of his opponents) culminated when, one fine day, he publicly burned the textbooks of medicine considered the gold standard at the time (i.e., the texts by the Greek physician, Galenus, who lived more than a thousand years before Paracelsus, or the Persian physician and polymath, Ibn Sin, also called Avicenna, who lived some 500 years before). Paracelsus left the university two years later and continued his travels. He died at the age of forty-eight in Salzburg (probably from the toxic effects of mercury).

A Novel Approach

Paracelsus is often referred to as the "Founder of Toxicology" or, if you will, the first clinical pharmacologist. (If you're a toxicology or pharmacology student, I'm kicking at an open door—there's no way around him: first year, first lecture, within the first ten minutes, his name will likely pop up.)

At the time, Paracelsus had revolutionized the field of medicine in a number of ways, and many of the basic principles are still valid today (with slight modifications and updates, of course). Here are a few highlights.

First, he advocated the use of plants or chemicals to treat diseases (instead of surgical procedures like amputating limbs)—leading the way to *phytotherapy* and *therapeutic drugs*.

Second, he suggested that disease located in a specific organ could be treated with agents that would exert their greatest effects in that particular organ—a forerunner of *target organ toxicology*.

Third, he encouraged the use of animals for the study of a chemical's effect and toxicity—building the foundation for *experimental pharmacology and toxicology*.

Fourth, he believed that a poisonous compound is amplified in a person with a poor underlying general condition—setting the stage for the principles of *genetic predisposition* and *idiosyncratic toxicity* (to be discussed later in this book).

Finally, and importantly, Paracelsus is most famous for his notion that "the dose makes the poison"—introducing the concept of the *dose-response*, one of the cornerstones of modern toxicology.

What an astonishing list of concepts for a time when the traditional opinion of many doctors still was to "treat" a wound with cow dung or bird feathers.

The Dose Makes the Poison

For freaks of 16th century-German, here is the full version of that revolutionary dogma by Paracelsus (published in the *Third Defense of the Writing of New Prescriptions, Basel, 1538*):

> *"alle ding sind gifft und nichts ohn gifft/allein die dosis macht das ein ding kein gifft ist"*

which means, *everything is a poison, and nothing is without poison; the dose alone makes that a thing is not a poison.*

Before we move on to discuss the far-reaching consequences of this statement, but also put it into perspective and relativize it a bit, we must have a closer look at what exactly "dose" means.

The Dose-Response

In toxicology and pharmacology, the *dose* is defined as a measured amount of a chemical that is ingested, inhaled, gets on our skin, or is being injected into our body. The term is perhaps best known from prescription drugs where the dosage of a certain medication is stipulated—how many pills per day, and for what duration of time. ("Dose" is sometimes used synonymic with "exposure," which can be confusing and inaccurate—this will be discussed later in this chapter.)

Everybody knows that a drug's effect on a patient can greatly vary depending on the dose—a drug may be useless if taken at a dose that's too small. It will only do its desired job if taken at the prescribed therapeutic dosage and may cause adverse effects if taken at an exceedingly high dose. The duration is equally important—some drugs may cause cumulative effects when taken repeatedly, or, quite the opposite, gradually lose efficacy when the same dose is repeatedly taken for a prolonged time.

Harking back to the above example of the chili peppers, increasing doses of capsaicin will induce increasing sensations of lust, then pain, finally agony, and this relationship is called the *dose-response*.

In its simplest form, a dose-response curve may be linear (i.e., with increasing dose the effect will always increase by the same degree (see Figure 3.2A). However, in real life, dose-response curves typically exhibit another shape: starting out flat, then exponentially increasing, and then flattening out again, reaching a plateau; because it resembles an S, this is called a sigmoidal dose-response curve (B). Naturally, to get the full picture for a given chemical, looking at the dose-response curve is more meaningful than just looking at a single dose point.

This may look like dry theory (where is the everyday application?) but dose-response relationships are essential in understanding the toxicity of chemicals. Therefore, let's consider another important point.

For certain chemicals, when you are increasing the dose, you don't see an immediate toxic response. In fact, you may have to initially crank

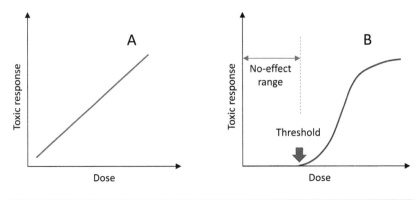

Figure 3.2   Shapes of typical dose-response curves. A, linear; B, sigmoidal curve.

up the dose quite a bit until you will see an effect (see Figure 3.2B)—in other words, there is a *threshold*. Beneath that threshold dose, the toxic response does not occur, and you have a "safe" no-effect range. (For lovers of details—maybe there is a minute effect, but you are unable to detect or observe it. So, it would be more accurate to talk about a *no-observed-effect* range.)

Imagine you are investigating whether a certain chemical in our environment might cause liver tumors in lab rats. You plot the dose-response curve from that animal study (because for obvious reasons you don't have the data for humans), and you observe a distinct no-effect window up to a threshold dose—what does it mean? It means that that chemical does not induce liver tumors in rats at doses smaller than the threshold dose. (If you wonder how these findings from lab rats could be translated to humans, you will find more information on how to do that in Chapter 15).

Back to Paracelsus's dose-dependence concept. The higher the dose, the greater the toxic effect—this makes sense, and it seems trivial to us nowadays (but it was not 500 years ago). But there is a caveat: does everything that we ingest actually get inside our body? The answer may be surprising.

The Dose Does Not Always Reach Its Target

When a hearty dose of a chemical agent is swallowed, inhaled, or comes into contact with our skin, it's not yet inside our body. There are barriers; e.g., a chemical must be absorbed by the gastrointestinal tract in order to reach the bloodstream where it then can be distributed throughout the body. If you swallow a pill that is poorly absorbed in the gut, most of that drug will not reach the general circulation, and therefore cannot be delivered to its target, e.g., the brain— probably not a good idea to market that drug as a wonderful antidepressant. The same thing holds true for the skin; most likely, only a small fraction of the dose on your skin will be absorbed across the layers of skin tissue and eventually reach the bloodstream (also called systemic circulation; *systemic* meaning the vascular system supplying the entire body with blood).

Time to slightly modify the Paracelsus dogma: instead of "the dose makes the poison," we could perhaps say, "the dose that is taken up in our body (that is available), makes the poison."

There is a term for that available fraction of the dose of a chemical that reaches the bloodstream: *bioavailability*. For example, 100% bioavailability means, all of the orally ingested chemical is absorbed, and the same amount will get into circulation as if you had injected the same dose directly into a vein (intravenously, i.v.). Only 10% bioavailability means, most of that chemical (90%) is not absorbed. (For interested parties who prefer a practical approach: how can a researcher determine the bioavailability of a chemical, e.g., a drug? One needs to take serial blood samples and compare the drug's concentration in the blood after oral administration with that of the same dose administered by intravenous injection.)

Figure 3.3 may help you understand the concept. At the time of injection of the drug into the bloodstream (time zero), you have the highest concentration, but then the levels of the drug gradually diminish (red curve), for two simple reasons: the drug is distributed to the different organs, and it is also slowly being excreted (e.g., through the kidneys). In contrast, after oral administration, the shape of the curve is different (blue curve); because absorption in the gut is slow, the highest blood concentration of the drug will be reached somewhat later, after which the drug levels continue to fall.

How can one quantitatively compare the two curves with each other? Instead of comparing the concentrations at a given time point, one can take the total integrated amount over time, the so-called "*area*

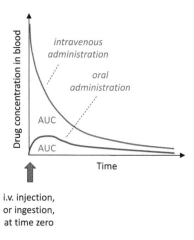

Figure 3.3   Typical area-under-the drug-concentration-versus-time curves (AUC).

under the curve," abbreviated *AUC*. (There are simple computer apps available to do such calculations.) For example, in Figure 3.3, the AUC after oral administration is quite a bit smaller than the AUC after intravenous administration; therefore, the bioavailability of that hypothetical chemical would be poor.

An example of poor oral bioavailability is mercury in its metallic form (as used in old thermometers). A single (acute) dose of accidentally ingested mercury does not pose a hazard because metallic mercury is poorly absorbed in the gastrointestinal tract. However, when exposed to mercury through a different route, i.e., by prolonged (chronic) inhalation of mercury vapors, the metal will be taken up via the lungs, posing a much higher risk for human health. (To preempt your possible objection: a totally different story is oral exposure to the highly toxic mercury salts or organic mercury compounds—but more on that is discussed in Chapter 13.)

So far, we've mostly talked about situations where we exactly know the dose (e.g., a pharmaceutical drug that we swallow, or a chemical that is administered to a rat in a clearly defined experiment). The situation gets a lot trickier when we are dealing with potentially harmful chemicals present in our environment—chemicals distributed in the water, foods, air, and soil, or concentrated at the workplace. Often, we can only make an educated guess about the dose we receive each day, let alone about the consequences. Sometimes we don't even have a clear idea about the overall exposure to those agents.

But wait a minute—what is the difference between dose and exposure? We will explore that in the next chapter.

TAKE-HOME MESSAGE

- Paracelsus, the founder of modern toxicology, has revolutionized the field by introducing a new concept: the toxicity of a chemical depends on the dose.
- The dose-response curve typically is sigmoid (S-shaped), sometimes having a threshold dose level with no apparent effect below that point, and often reaching a plateau at higher doses.
- Bioavailability is a measure of the fraction of the dose that reaches the systemic circulation.
- *Give-away*: A 500-year-old dogma can still be valid today.

4
EXPOSURE
The Key Determinant in Risk Assessment

What Does "Exposure" Mean?

The distinction between dose and exposure is quite finicky because the words can have different meanings in everyday language. In toxicology and pharmacology, *exposure* is the amount of a chemical in the environment (such as air, water, and food) that may come in contact with an organism (a quantitative term), but it can also simply be used in a descriptive way (e.g., the unprotected skin is being exposed to harmful UV). In contrast, the *dose* of a chemical is the amount delivered to the target (the body, a specific organ). Often, the terms are used interchangeably, but it may be better if we make the distinction throughout this book.

Apart from the question, "how much?" (referring to the dose), two other factors are also important for measuring the exposure (see Figure 4.1). It is the answer to the questions, "how long, and how many times?" (referring to the duration and frequency of exposure), and "how?" (referring to the route of exposure). We will discuss this in the next two sections.

Different Ports of Entry

When our body is exposed to foreign compounds (also called *xenobiotics*, as opposed to the chemical compounds of our own body), there are different front gates through which these chemicals can enter the body. The three major routes are oral exposure (by eating or drinking something), airway exposure (by inhalation), and dermal exposure (by putting something on the skin). For pharmaceutical drugs, the oral route no doubt is the most important one. Sometimes drugs are delivered to the body by intravenous or intramuscular injection which is not ideal for obvious reasons. These alternate routes are used when, for instance, the absorption of a drug from the gut is poor, or when the digestive enzymes in the

DOI: 10.1201/9781003346661-5

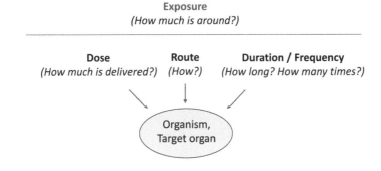

Figure 4.1   The three major variables of chemical exposure.

gastrointestinal tract would destroy the drug before it even reaches systemic circulation.

As to the oral route, there is also something unique about it. I'm not adding this to make things more complicated—I'm mentioning it because it's important: Any nutrients, but also foreign chemicals that are absorbed from the gastrointestinal tract do not directly merge into the systemic blood circulation. Rather, they have to pass a checkpoint first—the liver. And there's no way around it.

Checkpoint Liver

The liver has many tasks, but an important one is its function as an immigration checkpoint for everything that passes through it, in particular, those compounds that arrive from the gastrointestinal tract via the so-called *portal vein* (see Figure 4.2).

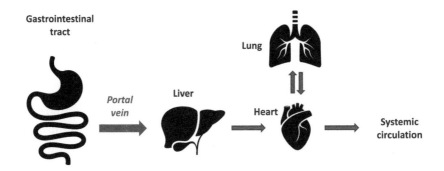

Figure 4.2   Role of the portal vein and liver in the systemic distribution of chemicals absorbed from the gut.

The portal vein collects the blood from the gastrointestinal tract, rerouting it to the liver before it can reach the systemic circulation (general bloodstream). Absorbed substances are therefore highly concentrated in the portal vein (the concentrations of some chemicals are up to fifty times higher in the portal than in the systemic circulation where they get diluted and distributed). This means that the liver is not only the first organ being "hit" by a potentially toxic compound after ingestion, but it also has to deal with a large amount of the chemical. However, the liver can do wonderful things: it can very selectively pick out a chemical and get rid of it (eliminate it through the bile), or, importantly, it can chemically modify that substance. Such a modified chemical is called a *metabolite* (much more on this is discussed in Chapter 6). A metabolite features new characteristics, different from the original chemical (the so-called *parent*). What does that mean for a potentially harmful chemical? In many cases, its toxic properties will be blunted. For pharmaceutical drugs this could mean that a drug can be chemically inactivated by the liver, thereby losing its efficacy long before it reaches the site in the body where it was intended to have its desired effect—something that a drug company is certainly trying to avoid. In other cases, the liver can turn a parent compound into a *more* toxic species; this will be covered later in this book.

Thus, it is only after running through the liver that the unchanged parent or the metabolites (there can be more than one) are released into the systemic circulation. If you start to feel that the liver may be some kind of a key player in toxicology, you're dead right!

How Much, How Often, for How Long?

The toxic response triggered by a xenobiotic not only depends on the dose and route (oral, dermal, inhalation) as we have learned, but also on the duration and the frequency of exposure. This implies that a hazardous chemical may induce a different type of toxicity after a short, one-time exposure as compared to that following repeated or long-term exposure.

A single dose of a potentially toxic chemical may be harmful if that dose is sufficiently high to induce an adverse effect. This is called *acute* toxicity. For example, an accidentally ingested household chemical, a pesticide spill, inhalation of a toxic gas like carbon monoxide, consumption of a poisonous mushroom, or over-dosing with a drug, all can trigger an acute toxic response. In fact, the example given in Chapter 2, where toxic agents were compared among each other with respect to their potency, referred to their acute toxicity.

While the information on the acute toxicity of a chemical is mostly gained from studies in laboratory rodents—actually, those data are required by the regulatory agencies for new products before they can be approved and marketed—the acute toxicity in humans becomes only known after anecdotal reports over time, e.g., after accidental exposure. When this occurs, a causal connection normally is obvious because the adverse effects arise soon and can be easily attributed to the causing agent.

However, there is an inherent problem with acute exposure to a chemical: as mentioned above, acute effects are not a good predictor of any long-term effects that could become apparent after prolonged or repeated exposure. For example, some people think when there is no immediate reaction (no acute toxicity) that the chemical must be safe. We've heard the hackneyed phrase: do you know anybody who has died after smoking one single cigarette or two?

If you still remember what a cigarette is—quitting wasn't that hard, right? After all, you've done it a dozen times—the toxic response of prolonged, repeated exposure to smoke (increased risk of lung cancer) can be quite different from that after the first acute exposure (cough-ing, nausea, wondering why people do this to themselves). Long-term exposure to a chemical is called *chronic* exposure. Not only can the type of toxicity greatly vary between acute and chronic exposure to an agent (with regard to different target organs affected, and different biomechanisms of toxicity), but chronic toxicity is also difficult to pre-dict based solely on the symptoms of acute toxicity. Unlike multiple exposures to radiation (or other DNA-damaging agents), for most chemicals, the resulting effects are not simply additive. For example, the symptoms after repeated exposure to a small dose of something (e.g., once every month for a full year) are not the same as those after receiving a single binge dose (the 12-fold dose once a year). To visual-ize this, just imagine if the effects of alcohol were additive, most of us would be permanently sloshed.

The crux with many environmental chemicals is that we may not know the human health risk they pose after chronic exposure (we're talking about years or decades). We simply have no clue, or we learn it too late. Another problem with chronic toxicity is that we're simply unable to avoid exposure to many agents in our environment. We can only strive at reducing the risk by trying to lower the exposure. The ALARA principle may be applied here, meaning *as low as reasonably achievable*.

To clear up any misunderstandings: chemicals that are newly introduced onto the market, like agrochemicals or pharmaceutical drugs in development, testing for chronic toxicity must be routinely done. Furthermore, lifelong exposure of lab rats to small, non-acutely toxic levels of new chemicals is strictly required by the regulatory agencies to assess their cancer-inducing potential; more on this is discussed in Chapter 15.

Exposure Determines the Risk

We have often heard arguments that a compound, say, a pesticide found on a produce, must cause harm—because the food analysts have identified it on those spinach leaves. While the concerns are justified and understandable, the quantitative aspect often goes unheeded. A chemical, even an agrochemical compound with high acute toxicity, does not necessarily cause harm *just because it is there*. It may be present at trace levels but can only be detected because modern analytical techniques have become so sensitive that they can pick up the minutest amounts (more on that is in Chapter 16).

Although the compound is there, the risk itself may be small (or negligible) for several reasons:

1. The exposure levels both for acute and chronic toxicity may be too low to be significant. In these cases, the dose may be below the threshold for inducing a detectable toxic response.
2. The bioavailability for that chemical may be poor, preventing it from reaching a target tissue.
3. There may be extensive chemical modification of that chemical and inactivation by the liver or by another organ before the harmful chemical reaches the target site.
4. Finally, the chemical may be rapidly eliminated.

$$\text{HAZARD} \quad \times \quad \text{EXPOSURE} \quad \Longrightarrow \quad \text{RISK}$$

Figure 4.3 The major determinants of risk.

(For the sake of clarity, there are a host of other possible reasons, e.g., our innate biological defense systems against toxic effects, but that will be covered in Chapter 8.)

It's all about exposure (see Figure 4.3).

Remember, even the most potent chemical is harmless when there is no exposure. That's why the handling of potentially dangerous household items, industrial products, or pesticides, must be done under strict safety precautions (e.g., wearing gloves, goggles, protective clothing, masks, or handling the chemical outdoors only)—with one aim: minimizing the contact with the chemical and thus keeping the exposure part in the equation as low as possible.

Now let's get a bit more practical—consider a nuts-and-bolts example that concerns us, consumers.

A Ubiquitous Metal

I am watching a rather boring popular consumer health series on TV when an announcement about the next upcoming topic startles me out of my doze: Aluminum and cancer. Aluminum in deodorants, to be specific. That sounds like an interesting headline.

I sit up straight, ignoring my beer and the pretzels, even my cell phone. I've heard about the issue before, in fact, wondered why in supermarkets many deodorants carry a supersized flashy label that says, ALUMINUM-FREE. Now a reporter is interviewing people on the street, asking them if they'd ever buy a deodorant that contains aluminum. Many of them couldn't care less, but those who are familiar with the topic back away, eyes wide as saucers, gasping for air, hands raised to ward off evil. "Never ever," they say. When asked why not, they say, unanimously, "because aluminum causes cancer." Quite an apodictic statement.

The reporter then decides to present the issue in the proper light and educates the viewers that back in the 1960s, when aluminum-containing antiperspirants and deodorant sprays became popular, the incidence of breast cancer in women was also increasing, and that, therefore, aluminum has been associated with causing cancer.

I can't quite follow that logic (you might as well blame loud rock music or flower power for that effect, or, on a more serious note, the introduction of the first hormone-based oral contraceptives that happened in the same period) but anyway. The show concludes by saying that we don't know enough about it and that the research is still ongoing.

With a grim sense of déjà vu, the single most important point that strikes me as I watch that educational program is that there has been no mentioning of any quantitative aspect whatsoever. No reference to hazard, exposure, dose, or risk. I'm left wondering why they don't explain the "dose thing" in simple terms.

So, let's get a few things straight.

Aluminum is one of the most abundant elements in the earth's crust. It is ubiquitously found in our environment. Therefore, humans have been exposed to aluminum (not aluminum metal itself, but aluminum salts) from natural sources in foods, drinking water, the soil, and plants. Aluminum has also been widely used as a food additive and is present in many cosmetics and pharmaceuticals (e.g., certain drugs that neutralize too much acid in the stomach to relieve heartburn and indigestion). True, we're getting exposed to aluminum daily, although to various extents. The average human exposure has been estimated to be up to 7 mg/person per day.

You know by now that this knowledge alone is not sufficient.

How about bioavailability?

Turns out that absorption of aluminum from the gastrointestinal tract is poor; measurements show values for oral bioavailability between 0.01 and 1.0% in humans.

As to absorption through the skin (relevant for our considerations about deodorants), the facts are even more sweeping. A recent clinical study showed that the bioavailability of aluminum from dermal exposure to antiperspirants was as low as 0.00052%. This and other studies clearly demonstrate that the amounts of aluminum that enters the body through the skin are too small to cause toxic effects in the body.

How about the hazard? Several studies demonstrate that repeated administration of high doses of aluminum (30 mg/kg body weight/day) to humans does not result in any significant toxic effects or accumulation in an organ. (Exceptions are high occupational exposure to aluminum or exposure in people with chronic kidney failure. Or, hypothetically, direct injection since then bioavailability would be 100%.) Also, aluminum has been associated with neurotoxicity in sensitive lab animal species, but the metal had to be infused directly into the bloodstream to elicit an effect. (For the sake of completeness, let's also briefly touch on the widely discussed potential link between aluminum and Alzheimer's disease in patients. The debate was stirred up when it was found that the brain of Alzheimer's patients exhibited excessively high aluminum levels—but it has been argued that this was merely a consequence rather than the cause of the disease. The debate is ongoing.)

Let's return to our initial question: how about an increased breast cancer risk? Chronic studies in mice with high daily doses of aluminum did not show an increased incidence of tumors. Taken together and based on all available studies in animals and epidemiological data in humans, it is not possible to conclude that aluminum is a potential carcinogen. There is no evidence showing that the use of antiperspirants increases the risk of breast cancer in women.

The use of aluminum in antiperspirants and other cosmetics is therefore considered "safe" for human health, as stated by the U.S. Food and Drug Administration (FDA) and European regulatory agencies.

Even if new methodologies in the future would allow a more accurate measure of the exact amount of aluminum absorbed from the skin and separate it from other exposure sites and sources (the background noise), there is no reason to panic, because there is no compelling evidence that shows that aluminum causes cancer.

Why is that not communicated to the public?

TAKE-HOME MESSAGE

- Exposure describes *how much* of a chemical is delivered (dose), *for how long* and *how often* (duration, frequency), and *how* (route).
- The liver plays a key role because it can chemically modify a compound after its entry and before it is released into the general circulation.

- Acute and chronic exposure to a chemical can cause different symptoms, severity, and target sites, and involve different biomechanisms. The toxic response to chronic exposure cannot be predicted from acute exposure data, and vice versa.
- The human health risk for a chemical is determined by its hazard multiplied by the exposure (or dose).
- Aluminum in our diet or deodorants, despite the metal's abundance and ubiquity, poses a low human health risk because both oral and dermal bioavailability is negligible.
- *On the side*: Be nice to your liver.

5

Natural and Synthetic Chemicals

Why That Chemophobia?

Some people believe that a lot of health problems can be chalked up to chemicals. While that may have some genuine truth in it, it's not what those folks really mean. What they mean is human-made chemicals (synthetics) that are added to foods or consumer products, as opposed to naturally occurring compounds—and here we are, right at the core of the misconception.

Fact is, *all* matter is made up of chemicals (atoms or molecules), everything we eat, breathe, and come in touch with. Our natural environment, even ourselves. (So, a "chemical-free drink" is not only a stupid marketing slogan but something that doesn't exist.)

Maybe that chemophobia is deeply rooted in associating the term "human-made chemicals" with negatively charged buzzwords like industrial pollution, harmful food additives, drugs with adverse effects, agricultural poisons, toxic spills, cancer-causing synthetics, and what have you. In the view of chemophobics, human-made chemicals are not only dangerous, but they also smack of something suspicious, incomplete, and artificial because they're not natural (and everything natural must be perfect, right?).

Well, not so fast. That belief could not be farther away from reality. Let's do a couple of fact checks.

"The most toxic chemicals are human-made." *Fact check*: Not true. The most toxic chemicals (with the highest potency) occur in nature, in animals, plants, and microorganisms.

"Each day we are exposed to many more artificial, synthetic compounds than to natural compounds." *Fact check*: Not true. It's the opposite.

"Only synthetic compounds can cause cancer (are carcinogenic)." *Fact check*: Not true. Among all those compounds that have been tested positive in rodent carcinogenicity testing, there is an equal number of natural compounds that are potentially carcinogenic.

DOI: 10.1201/9781003346661-6   **33**

Another interesting thing is that people often react, almost instinctively, with a dismissive attitude against something unknown that sounds, well, "chemical." The frequently cited joke about dihydrogen monoxide, which is corrosive to iron, and which can potentially kill people by suffocation, is an example. It's just, H_2O—plain old water.

So what's the difference between natural and synthetic chemicals?

Natural Versus Synthetic

A "natural" compound is a chemical that exists in nature (e.g., a certain metal), or that is formed spontaneously in the environment (e.g., ozone), or that is synthesized (biosynthesized) by a plant or an animal or a microorganism, without any human intervention. In contrast, a "synthetic" compound is one synthesized in a laboratory (sometimes called an "artificial" compound, like in "artificial sweetener," but I don't like that term because it bears a note of something fake, false, or simulated).

Let's take a simple example, like vitamin C (the popular term for ascorbic acid). Is the natural ascorbic acid (e.g., in a kiwi) the same as synthetic ascorbic acid (e.g., in a vitamin pill)? To make it, both the plant and the chemist use the same sugar, glucose, as starting material. The end product, ascorbic acid, is identical, regardless of its origin, and it works identical in the body. There is nothing bad about the synthetic version. (Of course, the vitamin pill that you pop every day may be covered with a harmless coating, protecting the cargo on the inside, but that's not the point here.)

Sometimes, though, things can get a trifle more complicated. For example, let's consider another widely used vitamin supplement, vitamin E. Naturally occurring vitamin E is found in many plant oils, seeds, and nuts. So, is the vitamin E produced in a lab chemically identical to the one found in extra virgin olive oil?

The naturally occurring "vitamin E" in fact comprises a small chemical family, a mix of eight different forms, including the most active member, alpha-*d*-tocopherol. In contrast, the synthetic form of vitamin E (present in skin lotions or in most dietary supplements) contains, besides alpha-*d*-tocopherol, alpha-*l*-tocopherol. This latter is chemically speaking the same, but, if you wish, is the mirror image of the former (like a glove for the right hand and one for the left hand). Why so? Because the lab chemist has used a different way to synthesize the compound than the plants do and so winds up with a side product that does not occur in nature. The biosynthesis processes in plants are sometimes asymmetrical.

Finicky? You're not a student of organic chemistry, so why bother? The example just helps to explain why the answer to the above question (is a compound exactly the same, regardless of its natural or human-made origin?) can sometimes be an unassertive "yes, but…. "

Obviously, the chemical industry does not only emulate and resynthesize chemical compounds that occur in nature. In the real world, a wealth of new chemicals is created by imaginative scientists, produced, and dumped onto the market each year—chemicals that to the best of our knowledge are not found in nature. Sometimes, a single new chemical is designed based on a rational, clever idea, in other cases, the quest for a suitable chemical simply follows a blind search for the hackneyed needle in the haystack. The sky is the limit for combinatorial chemistry (synthetic methods allowing to prepare tens of thousands of chemicals, followed by large-scale screening for the right one).

But oftentimes, the most amazing chemicals are found in nature.

Natural Medicines

Nature is a wonderful treasure chest for biologically active compounds including medicines. Our ancestors have known this for thousands of years. More recently, the pharma industry has been screening zillions of substances from exotic plants, fungi, or marine organisms, with the hope to identify new medicines (and they have found incredible stuff). Not because such nature-derived compounds are safe (only rigorous testing would demonstrate a new medicine's safety profile)—but because over millions of years organisms have developed fascinating compounds with astonishingly high potency. Some of them feature extremely high pharmacological activity. For example, they might fit snugly into a specific receptor in our body, like a key closely fitting into a lock, either triggering or blocking a biological response. Other natural compounds are highly toxic, but in tiny amounts could have a beneficial effect. And still others are natural antibiotics, natural painkillers, natural herbicides, and natural insecticides. (Think, e.g., of nicotine, cocaine, or pyrethrin that plants have been using as chemical-defense systems against predators.)

A great example of such a natural medicine is the serendipitous discovery of cyclosporine, a widely used immunosuppressive drug to prevent tissue rejection after an organ transplant. In 1970, a microbiologist from the Swiss pharma company, Sandoz (now Novartis), while on vacation in Norway, collected soil samples that would later be screened for biological activity. To their surprise, the researchers found that a fungus in those soil samples exhibited powerful immunosuppressive activity, which eventually led to the discovery and development of cyclosporine as a potent therapeutic.

Knowing that many naturally occurring chemicals are highly potent remedies or pesticides, exhibiting stunning biological activity, why don't we use exclusively natural products? What is the advantage of adding synthetic ones?

The answer is twofold: synthetics provide a broader variety of new chemicals with novel applications. Importantly, though, synthetic chemicals can be produced in bulky amounts, and they are often easier and cheaper to produce than extracting and purifying them from natural sources.

And something else: they are cleaner.

Natural remedies (e.g., extracts from plants) are often mixtures, containing a host of different unknown substances. Other characteristic differences are summarized in Table 5.1.

Therefore, one must make sure that "harmless" natural remedies don't have untoward effects on the body. Take, for instance, the plant extract from St. John's wort, which has been used therapeutically as an

Table 5.1 Characteristics of Natural and Human-Made Remedies

NATURAL REMEDIES ACQUIRED FROM PLANTS OR ANIMALS	SYNTHETIC (HUMAN-MADE) REMEDIES
• Often not tested for safety and efficacy • Extracts, mixtures, or semi-pure • May be contaminated (with viruses, heavy metals, toxic plants, etc.) • Content in plants may be subject to seasonal/regional differences	• Rigorously tested for safety and efficacy • Pure substance • Not contaminated • Highly standardized production

herbal antidepressant. The natural product seems to work fine, but in some cases, it was noted that the medicine interfered with the action of other drugs taken at the same time. For example, concomitant use of St. John's wort extracts and cyclosporine (the immunosuppressive drug used to prevent organ transplant rejection; see above) resulted in medical complications, in the worst cases even transplant rejection. What was going on?

It turned out that in those patients the blood levels of cyclosporine were way too low despite receiving a regular dose so that the drug's beneficial effect was lost. Why? It was soon discovered that cyclosporine was metabolized and eliminated from the body at a much higher rate than it was supposed to. The culprit was St. John's wort. Now we know that an active component in the extract (the chemical, hyperforin) can crank up the liver's capacity to metabolize drugs like cyclosporine (and many others) so that they lose their therapeutic effect and are eliminated much faster.

Because many naturally occurring chemicals have a high potential for therapeutic application (Nature is incredibly ingenious), the pharmaceutical industry has taken advantage of that knowledge by trying to use the natural chemical as a starting point to develop new medicines. They chemically modify the original compound and screen the products derived from the original template, e.g., for better bioavailability, lower acute or chronic toxicity, or fewer adverse effects. This has led to natural-derived chemicals.

Of Natural and Organic Pesticides

Synthetic pesticides have been incriminated for posing a chronic health risk to consumers, and many of the really bad ones have long been banned (although they may still hang around in the environment). As discussed in Chapter 10, for other currently used pesticides the safety issue is still a matter of debate, and more long-term research is warranted.

A major concern about these pesticides has been their residues on foods and in the water, and the involved potential risks to human health. In recent years, increasing awareness, but also organic food production (without the use of synthetic pesticides) has indeed reduced the amounts of residual pesticides in our foodstuffs. However, we should not disregard that other serious health issues are associated with foods (including organic food) that may be of even wider concern for human health: one is foodborne illnesses (acute "food poisoning") from pathogenic bacteria (e.g., *Salmonella*), another one is residual antibiotics (that might promote the spread of antibiotic-resistant strains of pathogenic bacteria).

It is often stated that one way to get out of this quandary would be to use natural pesticides only. Would these natural pesticides be safer than synthetic ones, because they are of natural origin? It's not so simple. Let's consider two examples.

One of them is *rotenone*, a plant-derived chemical. Rotenone is a non-selective insecticide widely applied in households and agriculture. It has also been used for centuries to catch fish (therefore also called a piscicide, a fish-killing agent). (For lovers of details: if you wonder why rotenone is acutely toxic to fish but much less to mammals—it is easily taken up via the gills, but poorly absorbed through the skin or the human gastrointestinal tract.) However, its use in organic farming is no longer allowed in the United States and Europe. Once in systemic circulation, rotenone can easily cross the blood-brain barrier (see Chapter 8). In fact, in rats and mice, rotenone has been associated with the development of Parkinson's disease, and there is evidence from studies on farm workers that there is indeed a link between heavy use of rotenone and neurotoxicity.

Another natural pesticide is *copper sulfate*, which is widely used in organic wine production. Copper is the real culprit; it is a fungicide used to protect the vines from mildew and mold (sometimes even added to the wine before bottling). One of its problems is that being a chemical element (a heavy metal), copper cannot be degraded, so over time it accumulates in the soil and can damage the soil's microorganisms, ultimately threatening the vines.

Collectively, we have seen that both natural and synthetic pesticides can be harmful. But rather than playing both ends against the middle and demonizing one or the other category, a more rational approach would be to use those that pose the smallest risks and are environmentally sustainable, aiming at minimizing exposure without sacrificing the beneficial effects of pesticides in agriculture.

To wrap up, here's a stunning note about natural pesticides. Bruce Ames at the University of California, Berkeley, pointed out in a famous paper that 99.99% of pesticides that Americans consume in their diet are of natural origin—many of them even positive in lab rodent carcinogenicity tests!

High-Hazard Natural Toxins

As mentioned above, some of the most toxic chemicals (in terms of potency) are found in nature, in microorganisms, plants, and animals. They have likely evolved to afford some degree of protection in their environment or against predators. Make no mistake, tiny amounts (less than one microgram) of some of those toxins could kill a human being. In fact, botulinum toxin, a group of related toxins produced by the bacterium *Clostridium* is at the top of the list of all known high-hazard chemicals. A big reason for concern? Yes, of course, if you happen to fall victim to inadvertently ingesting some canned food that's been improperly prepared and infected with those bacteria, you may develop deadly paralysis. Also, some of these highly potent chemicals are extremely dangerous if they fall into the wrong hands. So, should we rank all these natural toxins as a top priority risk, are they major chemical threats to humankind?

The answer is no, and here are the reasons: the overall human health and environmental risk of these toxins are low because the probability of becoming exposed to such a toxin for most people is low. Cases are rare (although severe). The source of most toxins can be located, and exposure is avoided. But probably the single most important difference to the more significant human health risks is that those naturally occurring toxins are not omnipresent in the environment, that they are not persistent, and that they do not bioaccumulate in organisms (stay tuned for more).

An example of a natural toxin that is highly potent but poses a minor overall human health risk because of its sporadic and limited occurrence is the "paralytic shellfish poison." Certain single-celled marine algae produce a number of toxins including *saxitoxin*, which are potentially neurotoxic. Rapid algal blooms, usually due to terrestrial runoffs of fertilizers, sewage, or agricultural waste, can lead to an out-of-control growth and accumulation of these algae along the coastlines, giving the water a reddish color (hence the name, "red tide," although it has nothing to do with the tides). The toxins can be harmful to marine organisms (fish, dolphins, birds), but also for humans if people are exposed to the toxin, either on the skin or by inhaling the aerosolized particles. If not recognized, toxicity may occur in people who consume contaminated shellfish. Clams, mussels, oysters, and other shellfish can greatly accumulate the toxin due to their filter-feeding action. In extreme cases, the toxic response may be respiratory paralysis; half a milligram is lethal for a human.

Another example of a natural toxin is intriguing because it demonstrates that the borders between natural and human-made toxicants can sometimes become blurred. In the 1990s, strange mass mortality of bald eagles across several states in the United States created a stir. The large birds of prey exhibited poor coordination while walking or flying, and died from drowning, starvation, or injury. The search for the underlying cause was futile. It was only some 25+ years later that this environmental mystery could finally be unraveled: the culprit was a novel neurotoxin, produced by a specific form of "blue-green algae" (cyanobacteria). While elucidating the chemical structure of the neurotoxin, it was found that the molecule contained bromine, an element relatively rare in nature. Turned out that the source of bromine was a certain bromine-containing herbicide (diquat; see Chapter 6) that found its way into the lakes and that the blue-green algae used to make the toxin.

So far, we have learned that exposure is the key, making even high-hazard chemicals less risky when the exposure is low (and vice versa). Next, we will consider what happens to a chemical once a dose has reached our body.

TAKE-HOME MESSAGE

- Human-made (synthetic) chemicals have a bad reputation, whereas naturally occurring chemicals are often thought to be better, even harmless. This is a scientifically wrong perception and false premise.
- The most toxic compounds (highly potent, highly hazardous) occur in nature, in plants, animals, and microorganisms.

NATURAL AND SYNTHETIC CHEMICALS

- Plant-derived medicines are often extracts, mixtures of partly unknown compounds, with less rigorous testing for safety and efficacy than synthetically generated drugs.
- Some natural pesticides used in organic farming, can be toxic too. Irrespective of their origin, the least damaging and most environmentally sustaining compounds should be selected and applied.
- Despite the highly hazardous nature, the overall exposure of humans to natural toxins is limited, cases are rare and, in most cases, avoidable, and the chemicals are not omnipresent and do not accumulate in the environment or organisms.
- *Custom-built*: The difference between natural and artificial chemicals is more artificial than natural.

6

WHAT OUR BODY DOES
TO A CHEMICAL

Toxicokinetics

In this chapter we will explore what the body does to a chemical after being exposed to it. Once absorbed into the bloodstream, a foreign substance will be distributed throughout the body. It may be sequestered and stored in certain compartments for a short term (hours, days) or for a long time (months, years), depending on the nature of the chemical. Importantly, it can also be metabolized, i.e., chemically modified. Finally, the chemical will be eliminated and excreted from the body. The umbrella term for all these processes that alter the concentrations of a chemical in the body over time is *toxicokinetics,* comprising absorption, distribution, metabolism, and excretion, aka ADME.

Toxicokinetics is an extremely complex area, and it is beyond the scope of this overview to touch on everything, let alone discuss it in detail. I will just highlight a few key points (see Figure 6.1).

Circling back to Chapter 4, our body can take up chemicals via different routes: mostly by ingestion, inhalation, or through the skin. Once absorbed, the compound will reach systemic circulation (the blood plasma, i.e., the liquid component of the blood) from where it is distributed to all the organs. The rate of absorption differs greatly from one chemical to the other; e.g., some foreign compounds are so poorly absorbed from the gut that they don't enter the blood circulation—if that is a pharmaceutical drug, it couldn't be given orally but instead would have to be injected directly into the bloodstream.

If the absorbed chemical is water-soluble, transport in the blood is not a problem; however, if the chemical is fat-like and not very soluble in water (meaning, in the blood plasma), it is often bound to specific proteins that give the chemical a piggy-back ride to its destination.

If the chemical is absorbed in the gut, it will first reach the liver via the portal vein, as propounded in Chapter 4; otherwise, it will

DOI: 10.1201/9781003346661-7

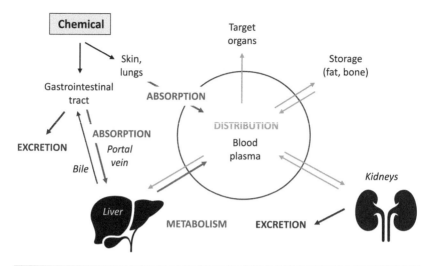

Figure 6.1 Uptake, metabolism, distribution, and excretion of chemicals in the body.

circumvent the liver and get into the circulation directly, from where it is distributed to all other organs or almost all, I should say. Some organs are shielded and protected against foreign chemicals better than others; stay tuned. Finally, a tissue or organ that is "hit" by a chemical that induces a toxic response is termed a target organ.

Cellular Transport Systems for Chemicals

Once inside a cell in our body, a chemical can easily move around. This holds true both for essential chemicals (e.g., energy-providing nutrients) and harmful foreign compounds. However, a chemical substance does not have unlimited access to all compartments in a cell. For example, the cell nucleus, containing the vulnerable DNA, will wall out most chemicals, not allowing them to enter unchecked unless needed.

But here's the point: how can a chemical get inside a cell in the first place? It's not so easy; cells are surrounded by a membrane (called plasma membrane or cell membrane), mostly made up of fat-like substances, studded with other supporting structures and specialized proteins. Chemicals on the outside of a cell cannot just walk across the plasma membrane, especially if they are bulky and water-soluble (and, therefore, colliding with the cell membrane). To get around this barrier, special molecular shuttle systems have evolved,

sitting in the cell membrane and allowing for efficient and fast transport of chemical compounds across the plasma membrane—not only
for import but also for export. Some of these transport systems are
highly selective, i.e., they only let certain specific chemicals through
while blocking out the rest; others are more promiscuous. How can
this be accomplished? Time for making a quick sortie—some more
details, I'm afraid but important to better understand how our body
handles chemicals.

The best chance to get across a cell membrane is for chemicals of
small molecular size and those that mix well with fats (so-called lipo-
philic chemicals). They can passively make their way across the lipid
membrane by a process called diffusion. The driving force for getting
across the membrane barrier is the concentration gradient (i.e., a large
amount of the chemical outside, a small amount inside). No energy is
needed for this "downhill flow."

Another, more efficient and more selective type of transport
involves a shuttle mechanism associated with the cell membrane that
greatly facilitates the influx rate—but only for certain chemicals.
This is accomplished by large, membrane-spanning proteins that
recognize and cherry-pick a specific chemical on one side, shove it
across the membrane, and release it on the other side. Such specialized transport proteins are called carriers (for larger molecules) or
channels (e.g., for sodium or potassium in nerve cells). Again, this
process is a passive downhill transport, driven by the difference in
concentration of the chemical on either side of the cell membrane
(going with the flow).

But what happens when a chemical must be imported or exported
by a cell *against* a concentration gradient? For such an "uphill transport" (swimming against the stream), simple carriers wouldn't work.
Instead, highly complex molecular "pumps" are in place that require
energy provided by the cell to move a chemical across the cell membrane. Again, we're talking about highly selective proteins built into
the membrane. It "makes sense" (to us) that these pumps are mostly
found in organs that have an important excretory function, such
as the liver, exporting chemicals into bile, and kidney, excreting
chemicals into urine. (For cell physiology nerds: the piling up of
chemicals on one side of the cell membrane by energy-consuming
pumps can sometimes be used to push a second chemical across the

membrane, even against a concentration gradient, via a co-transport or molecular "carpooling.")

Apart from their major role in the excretion of chemicals, such molecular pumps serve another important function: the concerted action of uptake transporters and export pumps limits the access of potentially damaging chemicals to vulnerable cells while at the same time allowing the uptake of necessary nutrients. Some of the most vital organs like brain, ovaries, and testes are protected in this way. This is accomplished either by tightly sealing the small blood vessels so that compounds are trapped, or by kicking out unwanted intruder molecules by a molecular bouncer at the entrance, if you wish—an export pump. In the case of the brain, this is the so-called *blood-brain barrier* (more on this later).

Collectively, what is allowed to enter a cell, how much of it, and where in the body plays a key role in the understanding of target exposure to harmful chemicals. To illustrate this, here is an example, again from the pesticide department.

Paraquat—A Hazardous Pesticide That Hijacks a Carrier in the Lung

Transmembrane carriers evolved to take up vital chemicals in cells and exclude others. As those transporters are quite selective, many foreign and potentially toxic chemicals are not readily taken up. However, there are exceptions; certain xenobiotics fit into the docking area of the carriers (like a key fitting into the keyhole) due to their specific size, shape, and other physicochemical properties, thus emulating an important chemical that the body needs. One example of such a toxic compound that gets a free ride is paraquat.

Paraquat is a non-selective herbicide (see Chapter 10) that was introduced in the 1960s and rapidly found worldwide application. After it was detected that paraquat can cause severe lung injury in humans (following accidental or suicidal ingestion), its use was banned in many countries, and strict tolerance limits were established. Unfortunately, other countries are still using paraquat, despite the human health risk.

After ingestion of paraquat, lung damage occurs slowly (over days or weeks), with fluid accumulation in the lungs, inflammation, and bleeding. Interestingly, it is the cells deep down in the small air sacs of the lungs, where the gas exchange between air and blood occurs, that are damaged. You might think that the damaging pesticide gets into these cells after inhalation, but that's not the case; instead, paraquat is taken up from the blood that streams into the lungs. How's that possible?

In these flat cells that line the cul-de-sac airways, the plasma membrane harbors specific carriers whose normal function is to take up polyamines (compounds needed for cell growth, cell division, and many other functions). Those airway cells have a much more efficient polyamine uptake system than other cells. Because paraquat shares certain structural features with the polyamines, the carrier cannot distinguish between the natural and the foreign "passengers" and shoves paraquat across the cell membrane. Once inside the cells, the high concentrations of paraquat can damage the cells. There are examples in other organs, too, for similar "Trojan horse-like" chemical assaults—sneaking in and doing harm inside the cells.

(For toxicokinetics enthusiasts: meanwhile, diquat, a structurally similar pesticide but one that has a greater safety margin, has vastly replaced its more hazardous cousin, paraquat. Interestingly, the diquat molecule does not fit snugly into the docking site of the polyamine carrier and therefore does not accumulate in those airway cells: no lung toxicity ensues.)

The Liver as the Major Metabolic Organ

When assessing the toxicity of a foreign chemical, it is extremely important to look at its *metabolism*, i.e., to what extent and how exactly the body will chemically alter the original chemical. In this context, the liver (again!) plays a prominent role.

The liver is often called the "chemical factory of the body" because it can metabolize many biomolecules produced naturally in the body, or essential nutrients taken up with food. Using similar processes, the liver has also the potential to chemically alter foreign chemicals, i.e., metabolize them into one or several derived new chemicals (*metabolites*). These modified chemicals have new properties, different from the original chemical (the so-called *parent*). For example, the metabolites are often more water-soluble than the parent and can therefore be more easily excreted by the kidneys.

In many cases, potentially toxic chemicals are thereby inactivated by the liver, "detoxified" if you wish (i.e., the metabolites are less harmful than the parent). In other cases, however, it can happen that such a newly generated metabolite is *more* toxic than the otherwise harmless parent—the liver has done itself a bad turn. In those instances when the generation of metabolites backfires, we talk about *bioactivation*.

Bioactivation of a harmless parent to a toxic metabolite is of paramount importance in toxicology. A widely known example is the painkiller drug, acetaminophen (in Europe and some other countries called paracetamol). At normal, therapeutic dosage, acetaminophen is safe because it is metabolized by the liver into several metabolites that are more water-soluble and therefore more easily excreted. However, after ingestion of excessively high doses, the liver generates a highly reactive, toxic metabolite that has the potential to kill liver cells across large areas, eventually causing liver failure. Unfortunately, this has been claiming the lives of many patients, mostly in the Western world—in the United States alone, we are talking about many thousands of emergency room visits and hospitalizations, and several hundred deaths from acetaminophen each year.

It should be noted, though, that the liver is not the only organ that can metabolize foreign compounds, although it is the most important one. Skin, lungs, or the intestines have similar capabilities. If you think of it, this biological function seems "plausible" (to us); it evolved in all these organs as they are the ports of entry where foreign chemicals come in touch with the body for the first time and need to be checked and, if necessary, inactivated and detoxified or, unfortunately, bioactivated and turned into more toxic species.

Another relevant example of a chemical that unfolds its toxicity through metabolic bioactivation is *benzopyrene*, one of several related products generated when organic material is burned. Benzopyrene is abundant in coal tar, in residential areas where wood is burnt, in cigarette smoke, and charbroiled meat, especially steaks with a generously charred crust. (For toxicology savants, there are several forms of benzopyrene; we're talking about benzo[a]pyrene, but for the sake of clarity let's stick with the simpler name.) Benzopyrene is poorly absorbed from the gut but easily absorbed in the lungs. The lungs, but also skin and liver cells, can bioactivate benzopyrene to a much more dangerous chemical species than the parent compound—in fact, activated benzopyrene can damage the DNA. The compound has been classified by the International Agency for Research on Cancer (IARC, an intergovernmental agency as part of the WHO) as a human carcinogen; lung cancer from cigarette smoke, or skin cancer from chronic exposure to soot, are well-known examples.

Gut Bacteria—More Than Just Quiet Coresidents

So, we've learned that the liver is the major chemical factory in the body and that other organs can also metabolize chemicals to a certain extent. We often forget, though, that there is another important site involved in metabolizing the body's own as well as foreign chemicals—the huge army of bacteria and other types of microorganisms residing in our gut. They are not pathogenic (not causing disease) but make up the perfectly normal and healthy gut flora called the *microbiota*.

The sheer quantitative aspect is overwhelming: trillions of bacteria normally reside in our intestine, and the total number of their genes (called the *microbiome*) greatly exceeds the number of genes of the human host. They can produce unique metabolites from chemicals—it's boggling, but beyond the scope of this book to summarize all their functions. Instead, let me give you one example to illustrate how normal gut bacteria can alter the exposure of a target organ (the intestine) to a pharmaceutical drug and in fact be involved in modulating its toxicity by metabolic activation.

Many of us should be familiar with a class of drugs called non-steroidal anti-inflammatory drugs (NSAIDs)—frequently prescribed and widely used painkillers (examples are diclofenac, ibuprofen, or indomethacin). What is also known about them is that prolonged use of high doses of NSAIDs may produce stomach ulcers in some patients; what is less well known is that some of them may damage the small intestine as well (an ulcer is a break in the intestinal inner surface layer, with loss of tissue due to cell death and bleeding, which becomes more dangerous as the lesions get deeper). We learned about this relatively frequent effect ever since it became possible to shoot pictures of the inside of the gut and its lesions by using miniature cameras that the patients swallow and that move along the gastrointestinal tract, providing factual proof.

Let's take a closer look at one of these NSAIDs, diclofenac. In an effort to get to the bottom of it, my former lab team at the University of Connecticut, as well as other groups, has been involved in this research, which explains my excitement. What happens when you let a mouse ingest a dose of the painkiller, diclofenac, at a relatively low dose? The drug is absorbed from the upper gastrointestinal tract, gets into the portal vein system, and reaches the liver. Specific enzymes in the liver convert the parent diclofenac to several metabolites; one major metabolite is generated by coupling the diclofenac molecule to a certain sugar—the new metabolite now is somewhat larger but much more water-soluble than the parent, allowing it to be more readily excreted. Subsequently, the diclofenac-sugar metabolite is "pumped" into bile, a ductular system that empties into the intestine. So far, so good. Normally, a metabolite in the intestine would be excreted in the feces—if it weren't for the hungry gut bacteria, waiting for new stuff to put their hands on. Bacteria have specific enzymes that can cleave off the sugar moiety from the harmless diclofenac-sugar metabolite (fair enough, the bacteria use the sugar as fuel for their own functions), leaving behind the original diclofenac, which is avidly taken up by the lower intestinal cells. Here, high concentrations of and repeated exposure to the drug can damage some of the cells, eventually leading to a thinning of the gut wall and the development of an ulcer.

Scientists are a curious species—their job is to ask questions. How do we know it is the gut bacteria that are responsible for activating a chemical and mediating the ulcers? Mice without any bacteria in their gut, or mice treated with antibiotics to deplete the intestines of bacteria, do not develop ulcers. Okay, but that's not yet convincing enough. You might argue that other, secondary factors may be responsible for triggering ulcer formation. A more direct approach would be to inhibit the bacteria's capability to cleave the drug-sugar complex while leaving everything else unchecked. But how could that be accomplished?

Matthew Redinbo and his team at the University of North Carolina have developed a specific and potent chemical inhibitor of that bacterial drug-metabolizing enzyme that splits the above-mentioned drug-sugar bond (the chemical inhibitor itself doesn't do any harm to the mouse cells and does not kill the bacteria). You guess where this is going. You treat a couple of mice with a high dose of diclofenac. Left untreated, they will develop ulcers. Only those mice who have also received the bacterial enzyme inhibitor are protected, they do not develop ulcers. Whether a similar strategy can eventually be applied to human patients will have to be determined in the future.

So, taken together, when talking about the metabolic activation of chemicals in our body, we should not forget to keep an eye on the gut microbiome.

The Kidney as the Major Excretory Organ

Most chemicals that enter the body do not stay there forever. Normally, chemicals and their metabolites are excreted, and their total amount in the body (the so-called *body burden*) is decreasing over time, provided the rate of renewed uptake does not exceed the rate at which the compound is excreted.

The two major routes of excretion of a chemical are via the kidneys (urine) or the liver (bile, which empties into the gut, the chemical being excreted in the feces). Most small and water-soluble chemicals or their metabolites are eliminated from the body in the urine, whereas larger and often fat-soluble compounds take the biliary route.

(For physiology freaks—an interesting biological fact is that the two excretion pathways accomplish their task through different mechanisms. In the kidneys, the blood is filtered passively and non-selectively, like a miniature sieve through which only the liquid component of the blood can pass, but not the blood cells and the large proteins in the blood. As a result, valuable nutrients, e.g., sugar or minerals, and lots of water would be lost in the urine—a wasteful process for the body. To avoid this, these important nutrients in the primary urine are being picked up in the kidneys, recycled, and transported back into the blood, while the products destined for excretion, including foreign chemicals, are eliminated in the urine. The liver, in contrast, works in an opposite manner: to get rid of chemicals or their metabolites, it selectively picks out what needs to be eliminated from the blood in the first place, and transports those chemicals into bile for excretion, while all the other substances remain in the blood circulation.)

There is a practical side to all this—the analysis of the urine, i.e., the quantitative determination of a given chemical or its metabolites in the urine, can be used to draw conclusions about the overall exposure of an individual to that chemical. The presence of such an excreted chemical or its metabolite is called a *biomarker*, an indicator of previous exposure. A typical example will be discussed in Chapter 9.

Hard-to-Get-Rid-of Chemicals

Not all chemicals that enter the body are getting metabolized. They are either eliminated as the parent or stored in the body, often for a long time. For example, certain fat-soluble chemicals can be stored in the fat tissue for decades. Among these are many slowly degradable pesticides (e.g., DDT) or dioxins. Other chemicals are stored in the bone; e.g., after chronic exposure to lead, most of the metal will be sequestered in the bone tissue.

If the rate of uptake and storage of a chemical in an organism greatly exceeds the rate of excretion, the net result is that the total amount of that chemical in the body will build up over time; in this case, we talk about *bioaccumulation*. If the hard-to-get-rid-of chemicals enter the food chain, and if their concentration increases with each step (e.g., from tiny plankton to small fish to larger fish to birds of prey), the process is called *biomagnification*. Specific examples of chemicals that undergo bioaccumulation and biomagnification will be highlighted in Chapter 13.

TAKE-HOME MESSAGE

- Toxicokinetics ("what the body does to a chemical") describes the changes in the concentrations of a chemical in our body over time, determined by absorption, distribution, metabolism, and excretion.
- Cells can selectively take up certain toxic chemicals by molecular transmembrane transporters or expel them

by molecular export pumps. This selective exposure explains why specific organs may be damaged while others may be left intact.

- The liver and other organs (e.g., skin, gut, lung) can metabolize, i.e., chemically transform, foreign compounds, facilitating their excretion, aka detoxication. In other cases, the liver can bioactivate chemicals to a metabolite that is more toxic than the parent compound.
- The gut microbiota (bacteria) is substantially involved in metabolizing chemicals.
- The major routes for excretion of absorbed chemicals are the kidneys (urine) and liver (bile-feces).
- Analysis of a chemical in the urine can be used as a biomarker of previous exposure of an individual to that chemical.
- If the rate of absorption is greater than the rate of excretion (elimination), a chemical may accumulate in the body.
- *Add-on*: Change is inevitable—for most chemicals in the body too.

7

WHAT A CHEMICAL
DOES TO OUR BODY

Toxicodynamics

In the previous chapter, we've learned what our body does to a chemical—here we will discuss the opposite: what a chemical does to the body. Obviously, a lot of good things, let's not forget that, from providing nutrients and energy to protecting the body and assisting it in fighting disease—but since our main topic is toxicology, we'll focus on the bad things.

Following exposure, and once ingested, inhaled, or taken up by the skin and "inside" the body, chemicals can do many different things— the way they exert their toxicity is called *toxicodynamics* (to set it apart from toxicokinetics).

So, what happens when a potentially toxic chemical hits a target, e.g., a specific organ, or a certain type of cell? It will interact with that target, either by chemically reacting with certain components of that cell in a random manner or, more specifically, by binding to a so-called "receptor." Receptors are large proteins with a highly selective docking area for certain chemicals only, debarring the rest of them. Ideally, the chemical will fit snugly into the receptor, like a key-lock pair, to again use that trite but vivid analogy. When this happens, the contact will trigger a specific toxic response (which, again, is dose-dependent). This response could send a signal to other cells or organs, which react in turn, amplifying or modifying the response.

(Before we move on to the nature of the toxic response, let's clarify something. We often tend to think of this sequence of signaling events in a linear fashion, like in that famous physics class experiment where the energy of the first ball in a row is transferred to the second, then the third one, etc. In reality, things are much more complicated, and the action-reaction branches out into different directions, forming a complex network of molecular, cellular, and organ-to-organ responses. Think of a snooker ball that hits a couple of other balls which in turn explode in all directions, hitting more balls.)

DOI: 10.1201/9781003346661-8

The molecular events, i.e., how exactly a chemical causes functional or structural changes in a cell, are called the *mechanisms* of the toxic response. For some chemicals, these mechanisms have been elucidated in the minutest detail, for most others, they are known in part, and for many others, they are completely unknown. Why is it important to study these mechanisms? you may ask. Wouldn't it be sufficient to just know a certain chemical causes, say, kidney cancer, and then just avoid that chemical? Why all these details? The simple answer is, the more we know about *how* a chemical causes an adverse reaction, the better we can gauge the human health risk. Here's an impromptu example—detailed studies in rats have revealed that certain chemicals abundant in unleaded gasoline trigger a toxic response, resulting in kidney cancer. Now, before you give all gas stations a wide berth and dig out your old bicycle, hold on a sec. Turned out that in order to produce kidney cancer these chemicals must bind to a specific receptor protein, and, fortunately, this receptor protein is present in male rats only, but not in humans. So, if you sell your gasoline-driven car for one or another reason, it shouldn't be because of a fear of getting kidney tumors.

It is beyond the scope of this book to discuss at length the numerous toxicodynamic pathways, let alone explain the complex biochemical and molecular mechanisms that trigger a toxic response. However, a few examples may help to illustrate the basic concepts. Among the many types of toxic responses caused by chemicals, we will briefly touch on some key functional and structural disturbances that can have dire consequences for the entire organism. Because this book talks a lot about pesticides and examines how dangerous some of them are in real life, let's look at four selected paradigms from the world of pesticides (but not limited to pesticides): neurotoxicity, hormonal imbalance, cell death, and cancer.

Targeting Nerve Cell Function

Some pesticides are potentially toxic to the nervous system (neurotoxic). In fact, some insecticides are specifically designed to kill insects by targeting their nervous system (stay tuned for more). The question is, are those chemicals selective for bugs or will they harm the nervous system of other animals too, including mammals and humans? Before answering this question, let's talk in more detail about one group of insecticides that has been hitting the headlines for years: organophosphates.

Organophosphates (don't ruminate on the chemical name) are a large class of chemicals used for plant protection, produced in huge quantities worldwide, and distributed in the environment. They have some advantageous features: once applied, they don't persist in the environment, nor do they accumulate in organisms. However, the big caveat is their high acute toxicity to the nervous system; human poisoning with organophosphates is often fatal. (Unfortunately, the knowledge about how organophosphates exert their neurotoxic effects has not only been applied for developing insecticides, but also for making chemical warfare agents, aka nerve gases.)

What happens on the toxicodynamic level? To understand how these hazardous chemicals act, we need a quick side-trip to recall how nerve cells transmit signals to various parts of the body, like muscles and glands. In a nutshell, an electric impulse coming from one nerve cell cannot simply jump to the next nerve cell, or a muscle cell. Instead, there is a chemical messenger bridging the gap between nerve cell 1 and nerve cell 2 (see Figure 7.1, left side). These messengers (called neurotransmitters) can vary; an important one in this context is *acetylcholine*. Acetylcholine is stored in tiny vesicles in the terminal branch of a nerve cell.

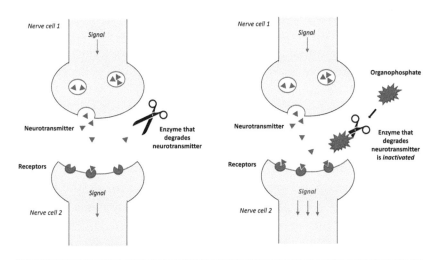

Figure 7.1   Normal function of neurotransmitter activating another nerve cell (left) and inhibition of this process by an organophosphate (right).

Upon a stimulus from upstream, i.e., when the nerve cell wants to "talk" to its neighbor cell, the neurotransmitter is released into the gap, from where it docks onto specific receptors on the surface of the nerve cell across. This will activate nerve cell 2, which in turn will conduct the signal further downstream, again as an electric impulse. Importantly, once the nerve signal has passed through, it must be discontinued; otherwise, the nerve cells would incessantly be firing away. To that end, there is an enzyme present that will degrade and inactivate the neurotransmitter. (For readers greedy for some biochemistry: the enzyme is called cholinesterase.)

Organophosphates can effectively screw up this crucial step in nerve cell function: they will react with the neurotransmitter-degrading enzyme, blocking its function. You can imagine what happens: instead of being removed, lots of the neurotransmitter molecules accumulate in the gap between the two nerve cells, where they continue to bind to the receptors on the other side. As a result, the overstimulated nerve cell 2 will send a much-exaggerated signal, and nothing will stop it (Figure 7.1, right side). The consequences of such an abnormally high and sustained presence of acetylcholine are huge—understandable because that part of our nervous system controls many different important body functions. In fact, acetylcholine receptor signaling is involved in the function of the intestine, glands, heart, respiratory system, and skeletal muscles; receptors are also present in the brain. Organophosphate toxicity, therefore, affects many vital functions, resulting in constriction of the airways, slowing of the pulse, salivation, diarrhea, muscle weakness, tremor, convulsions, and, in severe cases, respiratory depression and collapse of the cardiovascular system.

Time to tone down and clarify, before you turn away from consuming any agricultural products: because of the well-known, high acute health risk for humans, the producers of organophosphate insecticides have tailor-modified the chemicals to allow for more selective targeting of the insects' nervous system while largely sparing the human nervous system. Such pesticides, e.g., malathion, are not active from the beginning; they must first be chemically converted (metabolized) by insects to the hazardous product before becoming toxic and able to inhibit their neurotransmitter-inactivating enzyme. Humans, in contrast, metabolize the pesticide mostly through other pathways. But mind you, that news is not an all-clear call: embracing the possibility of repeating myself too often, very high exposure to these modified organophosphates can also become harmful to people. (On a side note: despite the potential toxicity of organophosphate agrochemicals, they cannot be directly compared with extremely toxic nerve gases. The potency of these latter differs from that of malathion by a factor of several thousand.)

After the nervous system, let's now briefly look at another important signaling system of the body that can be upended by certain pesticides and other compounds: hormones.

Disrupting the Endocrine System

In recent years, certain chemicals have fallen into disrepute for their potential to screw up our hormone balance (endocrine systems); they are therefore termed *endocrine disruptors*. For example, sunscreens have been decried for altering the sex hormone balance (stay tuned for more in Chapter 18); plasticizers, synthetic estrogens, anabolic steroids, and other compounds have been associated with harming normal development and reproduction. And one group of chemicals, in particular, has been brought into discredit: pesticides. We will discuss pesticides in much more detail in Chapter 10, so let's save the fun for later—only so much in advance: pesticides are a large group of diverse chemicals with very different characteristics, and their human health risks cannot be measured by the same yardstick. They cannot be vilified *en bloc* as being endocrine disruptors, nor, to be fair, should they all be trivialized as harmless substances.

Endocrine disruption remains a serious type of insidious toxicity that has gained enormous attention in the past years. Many mechanisms of the toxic response have been uncovered. However, one should be careful not to jump to premature conclusions from reports of *in vitro* studies only, i.e., acute studies performed on cell cultures exposed to exceedingly high concentrations of a suspicious chemical, as opposed to studies in live organisms. For a human health risk assessment, chronic low-level exposure data and dose-response considerations need to be factored in (much more of this is discussed in Chapter 17). But back to our topic—how can a chemical disrupt our hormone systems?

Hormones are chemical messengers produced at specific sites in the body. When released into the bloodstream, they are distributed across the body and exert an effect even at remote sites. Hormones are extremely potent; they will bind to and thus activate specific receptors present in certain cells and tissues. These activated receptors can then interact with the DNA and switch on certain genes (Figure 7.2, left side). Of special importance in this context are the sexual hormones (estrogens and androgens), but also other hormones including insulin and the thyroid hormone.

Certain foreign chemicals can interfere with the normal function of such hormone receptors. For example, endocrine-disrupting xenobiotics can emulate the natural hormones and thus disrupt the normal hormonal message pathway either by inducing or blocking an effect (see Figure 7.2, right side).

However, in most cases, but not always, those chemicals that fool a hormone receptor in this way are not as potent as the natural hormone. Higher concentrations and longer durations of exposure are required to elicit an endocrine-disrupting effect. Nevertheless, what has made some of the endocrine-disruptors notorious is their slow degradation rate and persistence in the environment, causing them to be a permanent potential threat. As mentioned earlier, some of them undergo bioaccumulation and biomagnification (a well-known example is the pesticide, DDT). Over time, their

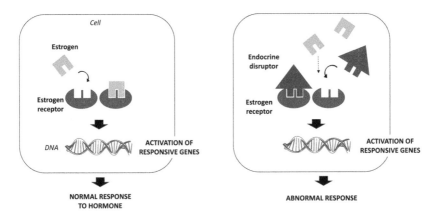

Figure 7.2   Normal function of estrogen receptor (left) and disruption of this function by an endocrine disruptor (right).

concentrations may get dangerously close to the tipping point where harmful effects are caused; in fact, in those cases, serious adverse effects can be triggered in susceptible species, including abnormal reproduction and development. In extreme cases, certain animal populations may decline, and entire ecosystems may be threatened.

So far, we've been talking about chemicals causing functional changes. Next, let's examine those dire conditions under which a chemical, without further ado, can directly kill cells.

Cell Death—By Accident or Suicide

Certain chemicals can injure cells to the extent that the cells will die. One example from the world of pesticides is methoxychlor (which will be revisited in Chapter 8). Cell death is the ultimate response when all compensatory mechanisms and rescue efforts of the cell fail. Cell demise typically occurs after exposure to high concentrations of a chemical and upon prolonged duration of exposure. How does the attacking chemical do that?

A chemical can strike out at one or several crucial constituents of a cell, e.g., the cell membrane that envelops the cell, or it can destroy the energy-producing machinery of a cell so that the energy supply will be cut off. The cell ultimately breaks down, and digestive enzymes are released from normally tightly sealed intracellular compartments, chopping up the debris, degrading the proteins, fats, membranes, and even the DNA. Typically, large areas of adjacent cells die in this way, and in the most extreme cases, the function of a targeted organ (e.g., the liver or kidneys) can no longer be maintained.

This type of cell death (by "accident") is called *necrotic* cell death. One example is liver necrosis associated with the painkiller drug acetaminophen. Echoing what we've learned in Chapter 6, a reactive metabolite of the drug (generated after a massive overdose) can trigger an avalanche of reactions, shutting off vital processes of the cell, eventually leading to necrosis in the liver. In fact, acetaminophen-induced liver injury currently is the single most important cause of acute liver failure in the western world.

Another type of chemical-induced cell death is cellular "suicide," called *apoptosis*. In contrast to chemicals that irreversibly damage and kill cells (see above), other chemicals can activate a cellular suicide program that is naturally present in a dormant state in all cells. Apoptosis is a complex process by which, once triggered, the components of the cell are surgically dissected so that the cell rapidly collapses like an erected tent whose supporting strings are cut. Typically, apoptosis occurs in single cells rather than across large areas, and the cellular remnants are rapidly and discreetly cleared away.

A well-known example of a chemical causing apoptosis is the plasticizer, DEHP, used in the plastic industry. The chemical is converted to a toxic metabolite (MEHP) that, at high concentrations, has been shown to trigger a massive wave of apoptosis in male germ cells, potentially leading to damage to the testes in young rats (more on plasticizers is discussed in Chapter 9).

The field of toxicodynamics is, of course, much wider than that illustrated by the few selected examples above. Chemicals can interact with the body and potentially do harm in many other ways, but a discussion of that would be way beyond the set framework for this book. However, to wrap up this chapter, let's talk about one type of chemical-induced disease that has attracted particularly wide public interest.

Cancer

One of the most frequently asked questions related to a hazardous chemical, including pesticides, is: *does it cause cancer?*

This wasn't always the case. For a long time, people believed cancer was a spontaneous disease, and nobody thought that chemical substances can cause cancer. This view changed when, in the late 18th century, the English surgeon, Percivall Pott observed that some of his male patients had developed skin cancer of the scrotum. As he learned that all these patients had worked as chimney sweeps when they were boys, he was the first to make an association between soot and cancer. (If you wonder why that particular area of the body was affected, other than because of neglected personal hygiene, here's why: the scrotal skin features a particularly high rate of absorption of chemicals.) It was only some 150 years later that it was understood how certain chemicals generated by combustion processes were causing the disease.

Nowadays we know that a host of chemicals have the potential to cause cancer (they are *carcinogenic*). The IARC (mentioned in Chapter 6) has established a long list of *human carcinogens* (see Appendix 2), and the list is growing. However, a better question than the one initially asked (does chemical X cause cancer—yes or no?) should be: does that chemical *increase the risk* of developing cancer when we are exposed to it via a specific route, at a given daily dose, for a certain duration of time? It's not about being punctilious, it's just scientifically more accurate (plus it will prevent a tsunami of anxiety-provoking thoughts). Again, what people should be considering is the risk, not only the hazard associated with a given exposure.

We've learned a lot about cancer risks and "risk factors" that contribute to cancer. A large amount of scientific data from combined epidemiological studies (studies in human populations), experimental studies in animals, and mechanistic considerations (chemical properties of compounds and their toxicokinetic behavior) have revealed that the greatest risk factors for developing cancer are unhealthy nutrition, smoking, and chronic infections, followed by workplace exposure, alcohol consumption, radiation, air pollution, and unknown "factors." Importantly, these data are average and apply to whole human *populations*; at the *individual* level, though, there are huge variations in susceptibility from person to person, caused by genetic factors, predisposition for certain types of cancer, but also by inadvertent exposure to different chemicals that, in many cases, goes unnoticed and is simply unavoidable.

However, we should not lose perspective and, instead, use common sense—here's an example. When you examine an alcoholic beverage, say, red wine, any analytical lab can likely identify traces of carcinogenic chemicals in it. If you're looking for them, you'll find them: lead, cadmium, arsenic, aflatoxin, among many others not mentioned in this book. But just because they are there, and because the current amazingly sensitive analytical techniques can pick up the tiniest quantities of different chemicals in a bottle of your favorite *grand cru*, these facts alone do not imply that the compounds are a major source of concern. Why not? Because, first, the amounts of those known carcinogens are extremely low and therefore of little concern (more on detection limits and safety margins later). And second, and most importantly, the real problem in alcoholic beverages is the alcohol itself. Alcohol (ethanol, and one of its metabolites) has long been recognized and classified as a human carcinogen—let alone the other human health and social aspects of excessive alcohol consumption.

Back to our topic. What is cancer, and how can chemicals cause it?

Cancer is the uncontrolled division of altered cells that invade neighboring or even remote tissues and organs. (*Tumors*, the umbrella term, includes both malignant tumors and benign tumors. Benign tumors don't invade other tissues but can be dangerous for other reasons.) There are many different types of cancer, in different organs, of different cellular origins, featuring different degrees of malignancy, but a detailed discussion is beyond the scope of this book.

For most cancers, it is almost impossible to retrospectively identify the cause. Adding to the confusion is the fact that cancer is not a one-hit-one-disease process; rather, it is a *multistep process* (see Figure 7.3).

The first step is the so-called "initiation" step in which a specific site in a gene of our DNA undergoes an alteration (a mutation). This can happen anywhere and anytime in the body; in fact, it happens more often than we might think. In most cases, though, the damage is repaired by a complex system of natural repair mechanisms. If such mutations occur in a non-critical region of the DNA, it is of little consequence. However, if the mutation occurs in a gene that helps regulate growth control of cells, and if this accidental alteration

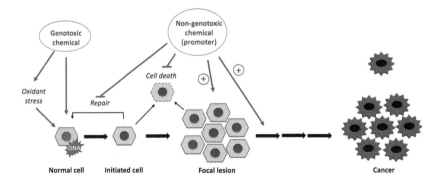

Figure 7.3   Chemical agents can initiate cells and/or promote the complex development of cancer at various stages.

escapes all the internal biological checks, this could pave the way for the next step.

The second series of steps is called "promotion." The mutated cell will divide and eventually form clusters of altered cells. More mutations may occur so that some of these altered cells may become potentially malignant, growing out of control. Finally, there is a third step, called "progression," which consists of multiple events that lead to the formation of a recognizable tumor.

Cancer is not only an astoundingly complex, but also a long process. The duration from the initiating event(s) to the diagnosis of a tumor can span over years or even decades, which explains why it is so difficult to pinpoint the cause. The incidence for someone to be diagnosed with cancer increases with age, at least for most cancers—it has even been claimed that everybody would eventually develop cancer if the lifespan of humans could be extended by many more decades.

So, where do chemicals come into play? Chemical carcinogens can act selectively at any one of these steps; they can be an initiator (by damaging the DNA), a promoter (e.g., by promoting cell growth or inhibiting the killing and removal of damaged cells), or even both. Accordingly, chemicals with a carcinogenic potential can be classified either as *genotoxic carcinogens* (initiators) or *non-genotoxic carcinogens* (promoters) (Figure 7.3).

Why is this distinction important? It may become relevant when trying to answer the question of whether there is a safe dose range below which exposure to a chemical may pose a negligible risk. For decades, scientists believed that there is no threshold, that even the smallest amounts of a genotoxic carcinogen could theoretically trigger cancer, while others held the view that there likely is a clear safe range at low exposure. It is beyond the scope of this short introduction to chemical-induced cancer to dive into the most current sophisticated mathematical models that discuss these events (we'll talk more on the practical aspects of cancer risk in Chapter 17)—just a glimpse ahead: it is conceivable, and it makes sense from what we've learned so far, that there are certain steps in the complex sequence of events that do have a threshold dose, especially for tumor promoters—think of detoxifying reactions that can become saturated when they are overwhelmed at higher exposure or protective compounds in our body that at a certain point become exhausted.

Persnickety? I'm afraid not—these considerations are not purely academic; they have major consequences for the regulatory agencies, putting them in a quandary. If there is no "threshold dose" for a chemical carcinogen, only a zero-exposure level would be 100% safe, but zero exposure is not possible in our world. Adhering to the no-threshold concept could also raise some visceral fear in people, increasing their—mostly unjustified—anxiety of getting cancer from drinking water or eating common foods. On the other hand, if there is an acceptable dose range, where should that magic limit between safe and no longer safe be set? (Stay tuned for Chapter 17 where we will discuss how a cancer risk assessment is done.)

Thus, we've learned that chemicals can cause all types of functional changes or structural damage to cells and organs. To re-apply our meanwhile trite tenet, it is the exposure that determines the type and extent of those toxic responses. But there's another factor that comes into play: our cells and organs are endowed with a host of natural defense mechanisms against chemical threats. Let's consider in the next chapter how these defense shields can dampen, or even prevent, a toxic insult.

TAKE-HOME MESSAGE

- Toxicodynamics investigates the interactions of chemicals with cells and organs and describes what types of toxic responses (functional changes, structural damage) they can elicit ("what a chemical does to the body").
- Neurotoxic chemicals interfere with peripheral nerve or brain function. Organophosphates (including some insecticides) block an enzyme that normally cleans up a neurotransmitter, resulting in overstimulation of many functions in the body mediated by the nervous system.
- Endocrine-disrupting chemicals (among them certain pesticides) can mimic or block hormones, potentially resulting in abnormal reproduction and development.

- The ultimate toxic response is cell death (necrosis), typically involving large areas in an organ. Chemicals can also induce cell suicide (apoptosis), triggering a cellular program for demise.
- Chemical carcinogens can be genotoxic (by damaging the DNA) or non-genotoxic carcinogens (e.g., promoting cell division of pre-cancer cells). While most non-genotoxic carcinogens have a dose range at low exposure without significant effects, it remains a matter of dispute whether there is a threshold for genotoxic carcinogens, making decisions about acceptable exposure limits difficult.
- *Supplement*: Shying away from indulging in a charbroiled steak once in a while because of its potential cancer-causing risk while at the same time smoking cigarettes and drinking booze is out of proportion.

8

DEFENSE SHIELDS

Standing Troops, Reserves, and Help from Outside

Each day we are exposed to a plethora of foreign chemicals, some of which are synthetic, human-made, and many others natural, but we seem to do quite well. "Of course," I hear you throw in, "because the dose is too small to elicit a toxic response." As explained in Chapter 6, our body can handle small quantities of those chemicals by metabolizing them, storing them away, and excreting them. But there are additional ways how our cells can defend themselves and avert potential harm by toxic substances.

When hit by certain foreign chemicals, our body has an astonishing ability to "sense" the presence of those chemicals and react to the impending threat of a potential "stress." For example, if the liver's capacity for metabolizing a given chemical is not sufficient to clear that compound away before it can accumulate and cause harm, the cells react by cranking up the production of specific enzymes involved in the metabolism of that particular chemical. As a result, the metabolic machinery works at full capacity now, and the equilibrium is restored, although at a higher niveau. This is called an *adaptive response* by the cells. (Apart from increasing the capacity of metabolizing enzymes, there are other types of adaptive responses, e.g., upscaling the antioxidant response—stay tuned.) However, please note that this does not work for every chemical and, importantly, that the adaptive response, preventing potential toxicity, is limited, and will eventually be overwhelmed as the exposure level keeps ratcheting up.

So, what happens when a hefty dose of a harmful chemical has invaded the body, posing a potential threat? Before cells or entire organs are being irrevocably damaged, there are a number of defense lines in place, and different kinds of troops to stop different invaders. To continue with our martial metaphor, some of them are standing army, ready all the time, to avert threatening danger: protective molecules present in all cells. Others are reserves, to be mobilized when there's an imminent strike: protective molecules that need to be made by our body, which may take some time.

DOI: 10.1201/9781003346661-9 **69**

On the public health level, say, in the case of acute poisonings, when help from inside the body isn't sufficient to prevent irreversible injury, help in the form of an antidote can come from outside. However, in contrast to popular belief, specific antidotes ("counter-poisons") are limited in number and efficacy.

Let's consider first the biological defense systems in our body that evolved to protect us from chemical insults.

Trapping Chemicals Before They Hit

Like an anti-ballistic missile defense system, the biological defense must inactivate an attacking chemical before it can strike and do harm. We have learned that chemically reactive metabolites generated in the liver and other organs are such potential threats (you may want to turn back the pages to Chapter 6). Among the body's arsenal of compounds to trap such reactive species is a ubiquitous chemical called *glutathione*.

Glutathione is a small peptide, made up of three amino acids; a mini-protein if you wish. It is much shorter than typical proteins, which are made up of several hundred amino acids in a row. In line with its role, glutathione is present at remarkably high concentrations in all cells. One of its major tasks is to catch reactive metabolites, chemically interacting with them, thereby defusing their harmful potential. If glutathione were not present, that reactive metabolite would likely react with a cellular protein and, when reaching critical levels, probably disable that protein's function. Worse even, certain reactive metabolites, unless captured, can react with DNA.

A perfect example to illustrate the protective role of glutathione is acetaminophen, the painkiller drug that we've encountered earlier. At therapeutic dosage, acetaminophen is metabolized in the liver into several harmless metabolites that are subsequently excreted. Small amounts of a chemically reactive metabolite may also be generated, but at this point, that's no big deal. There is sufficient glutathione present in the liver to intercept the dangerous metabolite before it can do any harm. However, if the safe daily dose is greatly exceeded, the glutathione stores get depleted, and it will take a while for the body to make new glutathione and replete the cells. In that case, the reactive acetaminophen metabolite has time and opportunity enough to

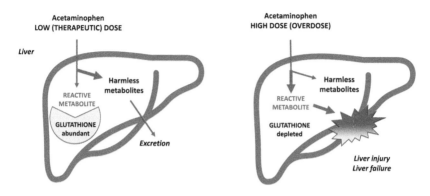

Figure 8.1 Glutathione protects against reactive metabolites of small doses of acetaminophen (left); large doses deplete the glutathione stores, and liver injury ensues (right).

attack cellular proteins and interfere with the cell's energy supply. As described above, this can result in necrotic cell death and, in extreme cases, liver failure (Figure 8.1).

(Anticipating the ensuing section on antidotes, here's a quick note on what the attending doctors will do in the event of an acute overdosing with acetaminophen: they will administer to the patient an antidote called acetylcysteine, which chemically resembles glutathione, and which is readily taken up by the liver, preventing further damage.)

Acetaminophen is a typical example of a drug that at a low dose is safe, thanks to an integral cellular defense system trapping and inactivating reactive metabolites. Another extremely important defense system in our cells and tissues is the antioxidant response.

Oxidant Stress, Radicals, and Antioxidants

Foods rich in antioxidants have been widely promoted as health-boosting agents, together with multivitamin supplements and exotic plant extracts (more on this is discussed in Chapter 12). However, apart from antioxidants that we supply from outside, the cells in our body are endowed by Nature with powerful inherent antioxidant defense systems. But what exactly are antioxidants? The name implies that they obviously counteract "oxidants," aka pro-oxidants.

Oxidative processes play a huge role in many human diseases, not just in toxic injury. Therefore, without wandering off the topic too much, it's time to define a couple of terms. I try to make it simple. (Readers with enough knowledge in chemistry should skip this page or else be prepared to cringe.)

When a foreign compound (in our case, the "perpetrator chemical") oxidizes another chemical, it means that it reacts with it and pries out a particle (an electron) of that chemical. The "victim chemical" is now a slightly compromised molecule, incomplete and imperfect, and in many cases unstable. It will desperately try to get that electron back from somewhere; it will likely hit another chemical in its neighborhood and in turn snatch an electron from that innocent bystander, starting a potential chain reaction. When large biologically important molecules in our cells are oxidized in this way, they might become more and more damaged over time, eventually losing their function.

That is the moment when an antioxidant comes into play. An *antioxidant* is a chemical that has the ability to provide that badly needed electron. It will restore the victim chemical by forcing an electron on it (the scientific term is, "reduce" it, which is the opposite of oxidizing it). The former victim chemical is complete again, but now the antioxidant itself has a problem: one of its own electrons is missing. Fortunately, there are processes in place to regenerate the antioxidant.

Electrons like to be in pairs. Chemicals that have either ceded or gained an electron in this take-and-give process (oxidation and reduction) and have an unpaired electron are called *radicals*. (The popular press, including health magazines, often talk about the bad "oxygen free radicals"; this is an outdated, incorrect term.) Radicals are often unstable and can launch a series of subsequent reactions, altering other biomolecules in the cell (imagine a chain of firefighting team members passing on the water bucket). Such radicals and other types of reactive chemicals that promote oxidations are called *pro-oxidants*. Usually, there is a balance between pro-oxidants and antioxidants, but when the antioxidant defense fails, or when the pro-oxidant attack is simply overwhelming, there is an imbalance in favor of the pro-oxidants, which is called *oxidant stress (or oxidative stress)*.

Foreign chemicals can greatly amplify oxidant stress (by complex mechanisms not discussed here), which, if left unattended, can lead to damage to proteins, fat, or even DNA, in our cells. In extreme cases, this can cause toxic injury. A grim example of such a chemical that intensifies oxidant stress is benzene.

Benzene is a widely used industrial product, used as a solvent and starting material for syntheses, and is a constituent of crude oil. In past years, though, its use in consumer products has been dramatically reduced because of its well-known toxicity. However, exposure to benzene cannot be totally avoided, as the chemical is present everywhere, including gasoline, wood smoke, and cigarette smoke. The bad news is that benzene has been classified as a human carcinogen—and unlike for other chemicals, nobody seems to dispute this fact.

What happens when we inhale or ingest benzene?

The liver metabolizes benzene, and the metabolites are distributed throughout the body. One of the metabolites has the potential to become a pro-oxidant. In most organs, this can be controlled and kept at bay by an efficient antioxidant defense mechanism. Unfortunately, one particular organ is less than optimally protected—the bone marrow, the place where the blood cells (red cells, leukocytes, others) are produced and later released into the bloodstream. Why is the bone marrow especially vulnerable to oxidant stress? For one thing, it is rich in an enzyme that can turn molecules like benzene metabolites into powerful pro-oxidants, and by the same token the bone marrow is very low on that antioxidant system that would counteract it—a bad combination, and the result is that the two processes keep goading each other. Indeed, benzene exposure in humans has been shown to be the cause of bone marrow toxicity—low blood cell counts and, in the most serious cases, leukemia (more on the human health risk of benzene is discussed in Chapter 13).

Looking back, it seems like oxidant stress, as well as other forms of cellular stress, is a bad thing. But, as we'll see in the following section, that's not always the case.

Stress Is Not Always Bad

Okay, we get the idea, every toxic response is dose-dependent. While greater exposure to a chemical (larger doses) may be harmful, lesser exposure (smaller doses) may produce milder effects. If the dose is below a certain threshold, the effects may even be silent, at least not readily detectable. But we can go even one step further and slightly modify this concept. In certain cases, a small dose of a potentially hazardous chemical may, in fact, be good for you—not only lacking an effect but producing a beneficial or protective effect that is greater than in the absence of that chemical. Contrary to what we've learned so far? Not necessarily.

If we look at an ordinary dose-response curve (see Chapter 3), the shape of that curve is sigmoid. However, in certain cases, the shape of the curve is biphasic (goes in two directions), i.e., with a bit of imagination, either resembles a "U" or an "inverted U," depending on what kind of response you measure (A or B, respectively, in Figure 8.2).

How is that possible?

All of us have experienced this phenomenon—called *hormesis*—in another context: psychological stress. Everybody is familiar with an exam situation, or what it feels like minutes before having to give a public speech or a musical performance. A little bit of acute stress is good as it keeps us sharp, awake, excited, improves our memory, and we feel the stimulating effect of that adrenaline rush. In contrast, too much stress, or prolonged stress, does the opposite; it's paralyzing, produces fear, freezing rational thinking. Another example is physical exercise: working out in a reasonable way is good for you, makes you fit and conditions your body; overdoing it not only gives you a charley horse but can also do more harm than good.

If we translate this to the cellular level, and assuming the stressor is a hazardous chemical, we can come up with the same concept. As

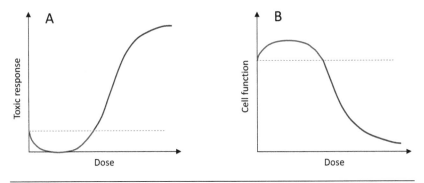

Figure 8.2 Examples for biphasic dose-response curves: hormesis. A, "U"-shaped curve; B, "inverted U"-shaped curve.

we've learned, chemicals can generate stress in cells or organs, e.g., oxidant stress, attacks on proteins, or putting stress on the metabolizing enzymes that can't keep up with detoxifying and exporting the foreign chemical. Too much of that stress would damage the function of that cell or even kill it. However, mild and transient stress can have the opposite effect: the cells can "sense" the stress and, as a response, specific genes involved in, say, boosting antioxidant defense are being turned on. As a result, cellular constituents would be better shielded from attack by freshly made molecular catchers that "quench" any threatening reactive metabolites. An example, again, is physical exercise, which initially induces mild oxidant stress but in the long run, is beneficial for one's health. Another example is the toxic metal, cadmium, which causes cells in the liver to rapidly make a protective chemical called metallothionein, which can trap the metal; stay tuned for more on that in Chapter 13.

In a nutshell, cells are highly adaptive and can activate compensatory mechanisms to fight off and counteract a threat posed by a foreign chemical.

Thus, low exposure to a potentially harmful chemical detected in our environment or food doesn't necessarily have to be dangerous; it could, in fact, strengthen the cells' defense forces. What doesn't kill us makes us stronger, is an old saying. However, we shouldn't miss the inflection point of the dose-response curve—after shooting over that beneficial dose level, things could soon turn ugly.

Refusing Unwanted Chemicals Admittance to Cells

Many cells are more resistant to toxic chemicals than one would expect. The reason is that certain compounds cannot reach the interior of the cell and stay there long enough to do harm in the first place. Instead, upon entry, they will be caught redhanded and kicked out.

Responsible for this selective deportation is a group of proteins associated with the cell membrane. These proteins are highly efficient; they have been compared to a "molecular vacuum cleaner" that can remove chemicals from the cell's outer membrane before they get a chance to advance toward the cell's interior. We have already encountered such export pumps in the section on toxicokinetics (Chapter 6). Similar pumps are present in organs that need to be specially protected, like the blood-brain barrier that we've introduced earlier.

An example to illustrate the effectiveness of such a protective molecular pump is, once more, a pesticide. *Methoxychlor* is an insecticide that has been banned in the United States and European Union in the early 2000s because of its toxicity and persistence in the environment (unfortunately, the insecticide is still being used in other regions of the world). Methoxychlor is an endocrine-disrupting agent (see Chapter 7), and exposure to high doses for prolonged time periods typically causes, among other bad things, damage to the male reproductive organs in lab animals. Because the testes, like the ovaries, need to be particularly shielded from the potential toxicity of foreign compounds—potential DNA damage could be transmitted to the next generation—the so-called blood-testis barrier must function as a critical checkpoint. Key to this barrier is a specific molecular pump, a protein that denies methoxychlor access to the testes; the protein is called MRP1.

To demonstrate the crucial role of MRP1 in protecting the male reproductive organs from methoxychlor-induced damage, let's briefly consider how this was done in practice. Lisa Bain and her team at the University of Texas at El Paso designed an elegant study in mice. When normal, healthy mice were exposed to a high dose of methoxychlor (25 mg/kg body weight) for 39 days (that's exactly the duration of a sperm-maturation cycle in mice), none of the mice had any significant changes in their testes. In contrast, when mice with genetically deleted MRP1 received the same treatment, they exhibited marked testicular toxicity, i.e., low numbers of developing sperm cells and lots of necrotic and apoptotic cells in the testes.

This and other studies have clearly demonstrated the importance of a natural, intact barrier system to protect organs from otherwise toxic chemicals. The next section deals with an outside intervention to protect the body against chemical toxicity.

Antidotes

An antidote is a therapeutic medicine given to a patient after exposure to a toxic dose of a chemical, in most cases after acute poisoning. A classic example is the oral administration of charcoal to mop up a toxic chemical that has been ingested. Another example is a therapeutic intervention to get rid of toxic metals; if their levels in the body are too high, the metals can be "flushed out" with a chemical compound that tightly binds the metal, and helps to eliminate it from the body.

In other cases, the approach is more specific: e.g., if a toxic chemical binds to a certain receptor protein in the cell where it induces a toxic response, the ideal antidote is a compound that fits even better into the receptor (to stay with the key-and-lock metaphor), displacing the toxic chemical. An example of this mechanism is the way an overdose with an opioid drug is being treated (see Chapter 14). The opioid binds to the opiate receptor, triggering well-known adverse effects including potentially fatal respiratory depression. The antidote of choice is naloxone, a drug that avidly binds to the opiate receptor, competing with the original opioid and pushing it away from the receptor, but without entailing the above-mentioned adverse reactions.

Unfortunately, for most chemicals, there is no specific antidote. And in the case of a slowly progressing, insidious, chronic toxicity, not much can be done, especially not when the toxic response is manifested months, years, or decades after the initial exposure.

TAKE-HOME MESSAGE

- The human body has different defense shields to ward off potentially toxic chemicals.
- Glutathione and related compounds can trap reactive metabolites and inactivate them.
- Antioxidants counteract the effects of pro-oxidants, compensating for the loss of electrons during oxidation. Oxidant stress is an imbalance between pro-oxidants and antioxidants in favor of the pro-oxidants.
- Small and transient exposure to certain hazardous chemicals that cause oxidant stress or other forms of cellular stress may even be beneficial, protecting cells by mobilizing the defense forces. The biphasic dose-response curve is called hormesis.

- Molecular efflux pumps in the cell membrane of brain, testis, or ovary cells can snatch foreign molecules and eject them from a cell before they can do any harm on the inside.
- Antidotes are medicines given after an acute intoxication; in general, they are of little importance after chronic exposure to an environmental chemical.
- *Encore*: Eliminating the primary source of damage is always better than constantly trying to clean up the mess—what works for a broken pipe in the house also works for fighting cellular stress.

9

CORRELATION AND
CAUSALITY

Does Drinking from a Plastic Bottle Cause Cardiovascular Disease?

One morning, over breakfast, the breaking news arrives like a bomb-shell. A hot media report in your favorite newspaper is conveying the shocking headline that certain synthetic chemicals that are widely present in a vast array of consumer products are linked with early death, prematurely sending tens of thousands of Americans to the grave each year. This is a bold statement, and you decide to continue reading.

The article talks about plasticizers, chemicals commonly added to a vast variety of industrial products (plastics) to improve their flexibility and durability. The bad, toxic chemical, you learn, turns out to be a group of so-called *phthalates* that are found in all kinds of tubing, food storage containers, cosmetics and shampoos, kid's toys, electronic products, automotive products, and other everyday items. Millions of pounds of these phthalate plasticizers are annually produced to improve the quality of numerous consumer products. Because the phthalates are just added but not chemically bound to the plastics, they become easily released and leak from those products—not sur-prisingly, they become widely distributed and are found everywhere in our environment, but, unfortunately, also in our bodies. They are virtually ubiquitous. There is no escape from being exposed to them.

You have never heard of these apparently highly toxic compounds before, and you decide to Google for the adverse health effects of phthalates. You find a host of information on the Internet—so it must be true, right? You learn that some of their untoward effects have been well-known for years, and that phthalates can cause a disruption of the hormonal balance and impair reproductive function, particularly in males. Widely documented, nothing to argue with that. Although you note, most of the studies were done with high doses and per-formed in laboratory animals. But what is new in today's media announcement, and quite disturbing, is that there is an apparent link between these plasticizers and cardiovascular toxicity in humans.

DOI: 10.1201/9781003346661-10

Hypertension? Clogged blood vessels? Coronary heart disease? Maybe even heart attacks? You stop reading, listlessly poke around the eggs and bacon on your breakfast plate and realize that you might have a slight pain in your chest when you really think about it long enough. Your appetite is gone. It's clear what must've caused it. You decide to avoid everything made of plastic from now on—but that's not only unpractical but downright impossible. And it's probably too late anyway. You're doomed.

Then, remembering the importance of the dose and exposure concept that we discussed earlier, you realize that there is no mention of a dose-response in that article. Without any quantitative data, it's difficult to gauge a human health risk.

If you have the moxie to dig into the original scientific literature, I must commend you—scientific papers are often hard to read, dry, complicated, and confusing not only for laypeople. What you would find in those international, peer-reviewed, high-ranking publications, surprisingly perhaps, are clear statements that there is insufficient evidence and lack of consistency of an association between exposure to phthalates and cardiovascular disease. So, according to current scientific knowledge, it seems far from being unequivocally proven that phthalates cause cardiac problems.

But then you remember that one can never "prove" something in science anyway, just provide evidence for or against something. You finish your cholesterol-laden breakfast and reach for a drink of water from the plastic bottle.

Is a Correlation Enough to Make a Strong Case?

That flashy newspaper article is still nagging at you, you can't help it. You wonder how it is possible to come up with such a bold statement without considering the numbers. Is there really a clear, believable association between phthalate exposure and cardiovascular disease? And how would the experts be able to demonstrate that such an association exists?

The first difficulty is that you would need to estimate the exposure of a person to phthalates over time—a task which doesn't seem to be easy. But the researchers can use a trick: because the phthalate plasticizers and their metabolites are mostly excreted in the urine, it makes sense to determine the different kinds of phthalates in urine samples and quantify them. Technically no problem, even for trace amounts; in fact, it's a daily routine. In this way, the urinary levels of phthalates and their metabolites would serve as an indirect measure, a biomarker (see Chapter 6), for previous exposure of a person to phthalates.

Haifeng Zhang, Xinli Li, and coworkers at Nanjing Medical University, China, analyzed data from urine samples of more than 10,000 individuals in the United States, and the different phthalates present in urine were determined by a highly sensitive, specific, and state-of-the-art analytical technique called *high-performance liquid chromatography-tandem mass spectrometry* (LC-MS/MS) (a crackjaw term which I'm simply mentioning because you might stumble across the abbreviation elsewhere). Now, in the same patients, the signs and symptoms of previous and current indicators of cardiovascular disease (e.g., high blood pressure, angina, heart attack, and stroke) were evaluated and their severity weighed. Next, the urinary levels of phthalates were plotted in a graph against the prevalence of cardiovascular disease, with many data points (each from a single person) through which a line can be drawn.

Other research groups have conducted similar studies. The results were clear: the higher the urinary concentration of a phthalate, the greater the severity of the cardiovascular disease. But does this association provide clear evidence that the phthalates actually *cause* the disease?

The obvious answer is no. What we have here is just a *positive correlation*.

Do Storks Deliver Babies?

You can "prove" anything by correlating two things. A widely cited, trivial example is the positive correlation between the reading skills of children and their shoe size. Nobody can argue with that. (On a more serious note, in Chapter 4 we've briefly discussed the apparently increased levels of aluminum in the brain of Alzheimer's disease patients—a positive correlation or a causative factor?)

You can even "prove" that the myth of storks delivering the babies to happy parents might have some truth in it. The White Stork, common in many parts of Europe, builds their nest on rooftops. Robert Matthews of Aston University in Birmingham, England, has come up with a half-serious but interesting study. Likely with tongue in cheek, he plotted the number of stork breeding pairs against human birth rates in several European countries and found, surprisingly, a positive correlation that was even significant when using sophisticated statistical tools. One of the logical explanations for the above-mentioned correlation might be that people in rural areas, where more storks are nesting, on average have more children than people living in urban areas.

As absurd as any overhasty conclusion may be, the example demonstrates the common fallacy of mistaking correlation for causation.

Plasticizers and Sex Hormones

Back to the cardiovascular toxicity allegedly induced by phthalates. Before jumping to a dismal conclusion ("plasticizers cause heart attacks"), we need to look at the big picture, trying to identify causative factors, not only relying on apparent correlations.

Phthalates have long been recognized as being a potentially serious human health problem. That is the reason why many of them have been banned from industrial production and use (e.g., in children's toys), and many others are tightly regulated in some countries. Since phthalates can make up to 50% of the total plastic mass (e.g., in PVC tubing), the overall amounts released into the environment are not peanuts.

As to the hazard, there are many different phthalates present in our environment. They belong to those chemicals collectively termed *endocrine disruptors*. Recalling the term from Chapter 7, this means that some of them have the potential to screw up our hormonal balance, in particular the sex hormones, causing problems with reproduction and development. Without demonizing the phthalates as an entity—we would need to carefully consider their exposure levels, on a case-by-case basis, for a founded assessment—phthalates clearly pose a red-flagged risk for humans. A host of studies suggests that exposure to phthalates could contribute to abnormal development in children, abnormal sex hormone regulation in women, and low sperm counts in men.

Besides alterations in hormonal balance, another effect associated with endocrine disruptors including phthalates is the development of *obesity, insulin resistance, and diabetes*. Obviously, there is a variety of risk factors for obesity, with endocrine disruptors being just one among many others.

It gets even more complicated.

As we all know, obesity and diabetes in turn are important risk factors for cardiovascular disease (stroke, heart attacks)—but they are not the only ones. It has been known for a long time that there are other predisposing factors for cardiovascular disease, including smoking, high cholesterol (LDL) levels, and high blood pressure. Could phthalates then just be another, indirectly acting, risk factor?

To say it more precisely, phthalates are likely involved in a complex network of causative factors ultimately leading to the development of cardiovascular disease (see Figure 9.1), rather than being a simple cause-effect pair. So, stating that plasticizers "cause" heart attacks, without giving the underlying context, is not only naïve, but also

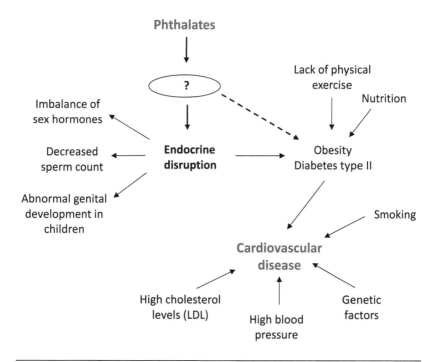

Figure 9.1 Phthalates and the complex network of risk factors involved in the precipitation of cardiovascular disease.

incorrect, unnecessarily ratcheting up the consumers' anxiety levels. It's not a relentless all-or-nothing thing; if phthalates were involved, they would at best *increase the risk* of developing heart problems.

On a practical level, how could the scientists provide more compelling evidence for a possible human health risk of the ever-present phthalates? First, we would need more epidemiological evidence to demonstrate causality between phthalates and human cardiovascular health risk. Second, we would need more experimental studies. Now, one might argue that many of the studies performed in laboratory animals or cell cultures might be questionable because of the difficulty to translate the results to humans. For example, how is it possible to study a chemical's effect on the complex cardiovascular system in a simple cell culture dish? Cell cultures obviously do not possess a heart, nor arteries and veins and lungs, and they do not build up blood pressure. Skipping to Chapter 15 for a moment, here's one very powerful advantage of choosing cell or organ cultures as an experimental model: the researchers can ask a very specific scientific question and then try to answer it by setting up a series of experiments with all necessary controls to minimize any confounding variables.

The goal of such cleverly designed experimental studies is to demonstrate *causality*.

The Search for Causality

There is never absolute proof of causality. In science, nothing is absolutely, 100% certain. But with clever experiments, one can provide compelling evidence for causation, so compelling that other researchers will believe it too. It is the study of cause and effect.

So, to stay with our example, after we've learned that there is a positive correlation between the urinary concentrations of phthalate metabolites and cardiac problems, is there enough believable evidence that phthalates actually contribute to *causing* cardiovascular disease in people? One theoretical experimental approach would be to put a total ban on the production and use of all phthalates (the cause) and wait until the environment is cleared from all traces of those plasticizers and then see whether the incidence of cardiovascular disease would be decreased (the effect). Needless to say, this kind of experiment, besides being naïve and unpractical, would also be inefficient. As we have seen, there are so many other risk factors in the complex network of events ultimately leading to, say, a heart attack, that cutting out one factor doesn't necessarily change a lot. What we'd need to do is using a reductionist approach, i.e., chop up the complicated sequence of events into smaller units that can be approached by experiments. Then you need to rephrase the question. The simpler the question, the easier it is to get an answer.

That's exactly what was done by a series of brilliant experiments.

Can Phthalates Cause Cells to Store Fat?

To understand the experiments, we must consider for a moment what's going on in our fat cells (adipose tissue). Inside these cells, there are several regulatory mechanisms that control the growth and survival of adipose cells and signals for them to accumulate fat. One of these molecular control switches is a protein called *PPAR-gamma*. When activated, PPAR-gamma can turn on certain genes that are involved in the handling of fat. Question is, what is the signal for this master switch to be activated?

You guess it: certain chemicals can switch on and trigger the process of fat production and storage. Béatrice Desvergne at the University of Lausanne, Switzerland, and her team asked the question whether a metabolite of one of the most widely used phthalate plasticizers (abbreviated as *MEHP*) can activate this signal protein, PPAR-gamma, and whether this would trigger the storage of fat in the cells. To get an answer, they used cell cultures of immature, lean fat cells and exposed them to MEHP—of course at levels that were realistic and in a similar concentration range as those found in human blood. They found that exposure of cells to MEHP indeed resulted in fat production and storage. But not enough—they could demonstrate that the signal protein, PPAR-gamma, must be present for fat accumulation to occur. When they blocked or deleted PPAR-gamma by a tricky molecular manipulation before adding MEHP, there was no response. Figure 9.2 (a painfully simplified summary of one of their experiments) illustrates the key experiment with its concurrent controls to demonstrate real causality.

What is the conclusion of that experimental study? We learn that the phthalate metabolite MEHP, at relevant concentrations,

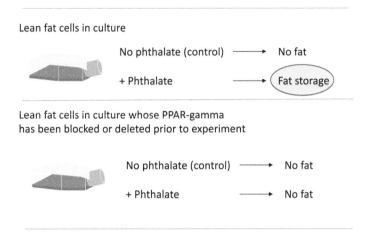

Lean fat cells in culture

No phthalate (control) ⟶ No fat

+ Phthalate ⟶ (Fat storage)

Lean fat cells in culture whose PPAR-gamma
has been blocked or deleted prior to experiment

No phthalate (control) ⟶ No fat

+ Phthalate ⟶ No fat

Figure 9.2   Experimental design to provide causal evidence of phthalates triggering storage of fat in cultured cells.

can signal the cells to produce and store fat. This is much more than just a correlation. At first appearance, this and similar other studies may seem isolated from the larger context, but—when looking at the bigger picture—could the results suggest that phthalates may be a contributing factor for the development of obesity, and perhaps even diabetes, in humans exposed to the plasticizers?

Clearly defined, "simple" experiments like the one described above (it wasn't actually that simple) are being used to convincingly demonstrate certain causal pathways. In general, individual experiments, dissected from the initial big problem into small chunks, allow for answering isolated but focused and pointed questions. It is only later that all this new knowledge can be assembled to better understand the overall, complex pathways of the toxicity of a chemical.

There is still a long way to go until we will exactly understand how phthalates and other endocrine disruptors can screw up the hormonal system and to what extent they might pose a relevant human health risk, including an increased risk for cardiovascular disease. Meanwhile, we cannot wait for unambiguous answers. Time is running out—and there will always be more questions, and the quest will never stop.

What we can do now is try to minimize the exposure to phthalates to levels low enough that they will not pose an unacceptable risk.

TAKE-HOME MESSAGE

- Do not confound a positive correlation with causation.
- Only rationally designed experiments with the proper control groups can provide compelling evidence for or against a hypothesis (one can never "prove" a hypothesis).
- Phthalates (plasticizers) are endocrine disruptors and just one potential risk factor for developing obesity and diabetes. They are indirectly linked to a complex network involved in cardiovascular disease. To say, plasticizers are "the" cause of cardiovascular disease is not appropriate.
- *Particularly today*: A clever and conclusive experiment has only one variable between experimental group and control group.

PART II
THE CHEMICALS

10

PESTICIDES

Killers with a License

Insidious Threat or Benefit for Humankind?

Imagine, just for a short moment, there are no pesticides.

> The human population is increasing, hunger is prevalent. The production of plants grown for human consumption is at risk because of an increasing rate of damage by weeds, rodents, insects, fungi, and plant diseases that all cause tremendous losses in crops. Half of the wheat produced is destroyed, the greatest part of the cotton production is gone, enormous losses are seen for soybeans, maize, rice, and potatoes. The changing climate will further increase the chances for draughts, flooding, prolonged heat waves, threatening agricultural production worldwide. In addition, mosquito-borne illnesses like malaria or dengue, are exploding...

These few lines may sound like a (less elegantly written but equally provocative) reverse image of the beginning of the famous book, *Silent Spring*, by Rachel Carson, published in 1962, which portrayed a planet without birds due to the detrimental effects of DDT, an insecticide lavishly used at the time because of the apparently negligible acute toxicity to humans. In case you didn't know: DDT has long been banned in the Western world but is still widely used in malaria-infested countries.

Pesticides, like most things including coins, have two sides. But why is it that in the past years, I've seen very few media reports that underscore the beneficial effects of pesticides on our society? Instead, the majority of popular articles have been focusing on the detrimental effects of pesticides, often on an emotional level, measuring everything by the same yardstick. All pesticides are not equal.

What I often miss, unfortunately, is a level-headed, rational, balanced discussion of the scientific facts, including quantitative exposure

DOI: 10.1201/9781003346661-12

data. So, for starters, let's define what pesticides are, and then look at the good, the bad, and the ugly side of them.

Pesticide Basics

Currently, there are more than 1,000 different pesticides known, acting by different biomechanisms. Each year, some 500,000 compounds are screened for their potential to become a new efficacious pesticide. Once discovered, they are not automatically given the benefit of the doubt—due to a history of unexpected adverse effects on the environment and human health, new pesticides undergo rigorous safety testing to avoid a repeat of the past (more on safety testing is discussed in Chapter 15). After all, pesticides are dangerous by definition, at least for pests.

Pesticide is an umbrella term—the three most important groups are insecticides, herbicides, and fungicides. (Since we're defining terms—often the expression *biocide* is used interchangeably with pesticide, but in fact, they are not the same. For example, biocides are primarily used for non-agriculture applications.)

Insecticides are designed to kill noxious insects; therefore, during the development of new insecticides, the aim is to select compounds that are toxic to insects but only minimally harmful to mammals and other animals. This is accomplished by developing chemicals that specifically target biochemical pathways important in insects, but ideally not present in other animals and humans. One of the most important (i.e., top-selling) insecticides is *imidacloprid*, belonging to the group of so-called *neonicotinoids*. "Neonics," by the way, apart from their major application in protecting seeds and foliage of plants, are also used to treat pets against fleas. Importantly, the neonics' harmful effects on the nervous system of insects are much greater than in mammals including humans; therefore, the acute toxicity of imidacloprid to humans does not seem to be a big concern. The compound is rapidly degraded by light and does not accumulate in the environment. Despite these positive traits, imidacloprid and some other neonics have been banned by the European Union in 2018—why? Because the insecticide has turned out to be a significant threat to useful insects, such as honeybees, reducing their populations and diminishing the biodiversity of wild bees that play an important role in ecosystems.

Next, *fungicides* are designed to kill parasitic fungi on plants and animals. One of the most widely used synthetic fungicides, although the compound has been initially derived from a mushroom, is *azoxystrobin*. It exhibits low toxicity to land animals including honeybees but may pose a risk to some water insects and fish.

Two widely used agents to inhibit fungal growths are sulfur and copper, both playing a role in wine production. Interestingly, these added chemicals have not received much attention from the public, possibly because they are "natural" fungicides and therefore believed to be harmless (a false doctrine, as elaborated in Chapter 5).

Finally, *herbicides* are designed to kill unwanted plants. Those used in agriculture are quite selective, i.e., they largely eliminate weeds while ideally leaving the crops intact. In contrast, other herbicides, including those used to clear roads and industrial sites, indiscriminately kill all plants they come in contact with. How about the human health risk? Some herbicides (e.g., the widely used *2,4-D*) resemble natural plant hormones and are therefore considered safe for animals and humans. For others (e.g., the synthetic paraquat), acute toxicity to humans has been a concern (we already talked about paraquat in Chapter 6).

Unfortunately, data on the chronic toxicity of herbicides are much harder to get than acute toxicity data. The gaps in our knowledge of their long-term adverse effects have fueled numerous discussions. For example, controversies still exist regarding the possible carcinogenic effect of certain herbicides in humans. Glyphosate is a typical example—stay tuned.

One of the most widely used herbicides is *mesotrione*. (For proponents of botanicals: I find the history of how that herbicide was detected quite intriguing. In contrast to the screening of thousands of chemical compounds for a possible herbicidal effect, researchers got a hint from nature. They had observed that the blossoms falling from the bottlebrush plant inhibited the growth of other plants in the surrounding areas, which prompted the isolation of a potent "natural" herbicide.)

Now let's take a closer look at a specific herbicide that has caused countless heads buzzing: *glyphosate*, and let's focus on the facts and figures.

Glyphosate: The Commotion

Glyphosate, the most widely used herbicide worldwide, has been restricted (or even banned) in more than twenty countries—so, surely it can't be entirely harmless, right?

That's a legitimate question. If that chemical were "entirely harmless," it would likely be biologically inactive and therefore a lousy pesticide. The problem lies elsewhere: many public discussions or flashy headlines about the safety of pesticides do not differentiate between hazard and risk (what else is new?)

To illustrate the point, here is an example, an excerpt from an imaginary round-table debate among four hypothetical good citizens (all profoundly convinced of the truth), followed by a quick fact check.

"Glyphosate is found on our food and is poisoning us every day," says the worried consumer. "Killing us slowly"—*Fact check*: Whoa, whoa, whoa... While small residues of glyphosate may be found in certain foods like pasta or cereals, the question must arise: how much? And will it harm us simply because it's there?

"Honestly, I don't understand all that fuss about it," retorts the carefree hobby gardener. "Glyphosate is absolutely safe for people. You can virtually drink it"—*Fact check*: Acute toxicity in humans is indeed rare and certainly not at the center of the pesticide controversy. However, there are isolated reports of people who developed a severe illness or died after accidentally ingesting large amounts of a glyphosate-containing pesticide. Not a common thing, though. What's of much greater concern is the chronic toxicity after long-term use of the herbicide.

"I heard that glyphosate is an organophosphate, related to the nerve gas agents, so it *must* be toxic," a visibly upset guy shouts, clearly wanting to impress the others with his chemistry knowledge—*Fact check*: Wrong. Glyphosate is a phosphonate, which is quite another cup of tea. Please don't confuse the two. (Some insecticides, e.g., malathion, do belong to the organophosphate class, but I promised in the Preface of this book that I won't dive into the chemistry, so I just leave it at that.)

"Everybody knows that glyphosate causes cancer," claims an overly zealous environmentalist. "That says it all"—*Fact check:* Not so fast. The statement is a shot from the hip and is not correct in this simplified form. We will discuss the issue below.

But first the facts. The toxicological literature about glyphosate is overwhelming, and steadily growing; here is just a very condensed and slightly digested summary.

Glyphosate: The Facts

Glyphosate is a broad-spectrum herbicide introduced on the market in 1974. Currently, the chemical is marketed under many different trade names and in different formulations. (The compound was first synthesized in 1950 by a Swiss chemist, but the work was never published. Think of all the money he could have made had he known the outcome... .)

How does this herbicide work? Glyphosate blocks a *plant-specific* pathway for the synthesis of certain amino acids (the building blocks of proteins)—which is great because animals would not be affected. The flipside is that glyphosate at high concentrations could also potentially damage crops, not only weeds. And that was exactly what happened, after the initial success, as some of the weeds spontaneously developed resistance to glyphosate over time, now having an advantage over more sensitive plants. To effectively eliminate the now partly resistant weeds, the amount of glyphosate applied to the crops had to be boosted—with the predictable result that the crops now became damaged too. A catch-22 situation?

The solution to the problem came when scientists introduced genetically engineered defense mechanisms into crops, making them both less sensitive to glyphosate and more efficient in breaking down the unwanted herbicide. As a result, as of 1996, corn, soybean, cotton, and other crops indeed became more resistant to glyphosate. The late John Casida of the University of California, Berkeley, one of the leading authorities on pesticides, called this "the most important of all applications of genetically modified organisms in crop production and the life sciences."

Unfortunately, a new problem emerged soon after the ever-greater amounts of glyphosate were applied: not only the weeds but also the genetically modified crops would potentially lose (at least part of) their resistance to the herbicide. One way to get around it was the addition of certain chemicals, so-called surfactants, during the commercial production of the herbicide. The beneficial effect was obvious: the added surfactants facilitated the uptake of glyphosate into the plants, making the glyphosate/surfactant mix more toxic to weeds than glyphosate alone. Unknown at the time, this had dire consequences for the overall safety of the new products, as we will see below.

Not enough—another challenge posed by raising the amounts of the herbicide applied to the fields was the unavoidable higher soil and water contamination and the possible impact on the environment. How bad is it really?

Fortunately, glyphosate for the most part has a favorable eco-toxicological profile. Simply put, it poses little harm to the environment. The chemical strongly binds to the soil where it becomes rapidly degraded by microorganisms so that it does not accumulate in the groundwater. Occasional surface runoffs may be a problem when high amounts of the herbicide reach rivers and lakes, but the good news, again, is that the toxicity to both aquatic and terrestrial animals is low (Figure 10.1).

(For eager students of eco-toxicology: it is easy to overlook certain things unless we ask a question from outside the box. For example, insects like bees have always been viewed as non-target organisms for glyphosate—after all, it's an herbicide, not an insecticide. But recent research by Anja Weidenmüller at the University of Konstanz, Germany, and her team suggests that glyphosate may have hard-to-spot effects on insects, such as behavioral changes in

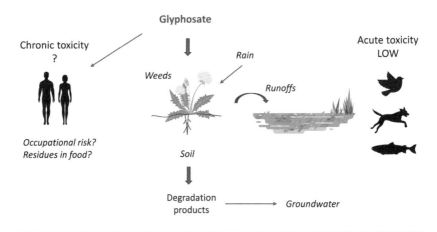

Figure 10.1   Environmental fate and potential health risk of glyphosate.

bumblebees (a surrogate species to study a large number of ecologically important wild bee species). Bumblebees fed sugared water containing glyphosate (5 mg/L, a concentration in the ballpark range found in nectar and pollen of glyphosate-treated plants) were unable to maintain the high colony temperature required for their reproduction.)

From eco-tox back to humans—what do we know about the potential toxic effects of glyphosate on people?

Acute toxicity to humans is negligible and may only be a problem at high doses, e.g., by improper handling. However, what is of much greater concern is the poorly predictable chronic toxicity due to sustained, daily exposure (Figure 10.1).

Long-term exposure to glyphosate has been associated with hormonal disruption and reproductive effects (harking back to Chapter 7, glyphosate is a potential endocrine disruptor). However, since many of the studies were performed on cell cultures or in laboratory animals, a direct one-to-one translation to humans is difficult. In addition, it turned out that the surfactant, added to glyphosate to increase the rate of the herbicide's uptake into plants, is toxic by itself, perhaps even potentiating the effects of the herbicide.

The greatest controversy, however, remains—might glyphosate be associated with cancer in humans?

Glyphosate: The Cancer Controversy

It often seems as if the glyphosate controversy was nothing but a battle between chemophobic environmental advocacy groups ("glyphosate must be banned because it causes cancer") and biased researchers of big agro ("there is no correlation between glyphosate and any carcinogenic risk in humans"). The way these exaggerated statements are worded, both are wrong. In reality, the situation is a trifle more complex.

Many studies on glyphosate were done *in vitro*, but we meanwhile know that what you find in a cell culture flask cannot simply be translated to the human situation. Again, it's the difference between an acute hazard (determined under very special laboratory conditions) versus chronic exposure (done under long-term, everyday-life conditions).

Epidemiological studies, i.e., studies trying to find a positive correlation between exposure to glyphosate and cancer in humans, have been equally tricky for a number of reasons (outlined in Chapters 7 and 17). One approach was to look at occupational exposure in workers who had been applying the herbicide. The result? No hard correlation with regard to cancer was found.

Nevertheless, in 2016, the IARC evaluated all available data on glyphosate and concluded that it should be classified as a *"probable human carcinogen"* (see Appendix 2). This was alarming, causing many countries to react and restrict or even ban glyphosate. In contrast, regulatory agencies including the US Environmental Protection Agency (EPA) as well as the European Chemicals Agency (ECHA), the European Food Safety Authority (EFSA), and the Joint FAO/WHO Meeting on Pesticide Residues (JMPR), who reviewed all the available evidence, concluded that judicious application of glyphosate *does not pose a carcinogenic risk to humans*. Which prompted the IARC two years later to reevaluate the issue without deviating from their initial conclusion. So, who's right?

The currently available evidence is simply not convincing enough to make a definite conclusion, and there is no consensus among the cancer agency, the researchers, and the regulatory agencies. If you wonder why not, here's one of the possible reasons: while it is easier to recognize that something happens, it is much more difficult, if not impossible, to say that it will *never* happen.

As to the other risks posed by long-term exposure to glyphosate, the endocrine-disrupting effects, the controversy is also ongoing. While regulatory agencies and a number of researchers conclude that there is no evidence of any endocrine-disrupting properties of glyphosate, others claim to have demonstrated effects on female and male reproductive hormone systems.

All this sobering uncertainty leaves us baffled. How should one act? Make decisions based on the hazard alone since the real risk is so difficult to assess. In the EU, the regulation on endocrine-disrupting chemicals generally is more hazard-based, while in the United States, decisions are clearly more risk-based.

What is the solution? Until we know more about glyphosate (and this holds true for pesticides in general), it is probably best to reduce exposure wherever possible. But let's get real—to ban the use of pesticides altogether would have enormous repercussions. The world's growing population needs food and clothes, and, with the changing climate, enormous challenges are heading our way. However, the "ideal" pesticide—high potency, high selectivity for certain weeds but leaving the crops unharmed, rapid degradation in the soil, and lack of toxicity to aquatic and terrestrial organisms including humans— doesn't exist yet.

Does Glyphosate End Up on Our Food?

What does all this mean for the average consumer? How about the alleged slow poisoning by pesticides via the food we eat and the water we drink?

Should you decide to dig into the original literature, you will find that glyphosate is indeed present on our daily foodstuffs. But wait a sec—if you belong to the few who've been brave enough to read this book up to this page, those grim findings shouldn't be much of a surprise to you.

The key question is, of course: how much?

Let's take a closer look. I am just citing one example representative of many others: in 2018, Otmar Zoller and colleagues of the Swiss Federal Food Safety and Veterinary Office analyzed different foods (not chosen randomly, but those that were suspected to contain residues of glyphosate) and found average levels of glyphosate ranging from 0.0005 to 0.1733 mg/kg food (very rarely, higher values were found). What does it mean? These numbers alone don't tell us much, so let's put them into perspective.

The Food and Agriculture Organization of the United Nations (FAO) has issued recommendations for maximum residue limits (MRL, see Chapter 16) that differ for the different food types. For example, the MRL for glyphosate is 1 mg/kg for nuts and 30 mg/kg for cereals. In view of these international limits, the amounts of glyphosate found in the study described above were way below the MRL. Therefore, the conclusion seems straightforward: based on rational reasoning and the current knowledge, there is little concern for human health risks.

TAKE-HOME MESSAGE

- Pesticides are *designed* to kill (insects, plants, fungi, etc.); therefore, they are biologically active, and their hazard and risk for both the environment and human health must be carefully evaluated.
- Qualitative data on the hazard alone are insufficient to characterize a human health risk; *quantitative* data are required.
- It is not possible to estimate the risk of chronic, long-term exposure to a chemical solely based on the available acute toxicity data. The outcome may be completely different between the two.
- Glyphosate (an herbicide, used as an example) exhibits low toxicity for the environment and humans. More research on a possible (low) carcinogenic risk of glyphosate is needed.
- A chemical does not necessarily cause harm just because traces of it are found on food and in the water. Even long-term intake of glyphosate residues on food is not a public health concern.
- *As an extra*: If you make an exciting scientific discovery, don't let your hot data go stale in a drawer.

11
Toxic Food

Ingredients, Additives, and Contaminants

An everyday scene, unfolding in the break room.

"Disgusting," one of your colleagues says, glancing at the raspberry yogurt in your hands. "And you're gonna *eat* that, seriously?"

There he goes again. Over time, you've become quite familiar with his concerns. "Why not? Anything wrong?" You can hear a twinge of annoyance in your own voice.

"Just read the small print." He jabs a finger at the plastic cup. "Food colors, antioxidants, flavor enhancers, stabilizers, thickeners, and an artificial sweetener. Yuck! Any room left for some natural ingredients?"

Usually, it's pesticide residues on veggies, carcinogens on broiled meat, all that bad stuff in the drinking water, that ticks him off. Today, it seems to be food additives.

"We got enough problems dealing with toxic residues you cannot avoid," he continues, "so at least the artificial stuff that they add to the food should be safe."

"It *is* safe. And, by the way, don't lose sight of the *real* concerns of food safety," you say, your tone leaving no room for arguments. "When it comes to food-borne risks, by far the greatest risk threatening us consumers is contamination with microorganisms. Food poisoning is what sends us to the hospital, not some harmless chemicals. Toxins, produced by bacteria or molds."

"Thanks for the lecture. I prefer chemical-free food."

You shrug in surrender. What can you say?

In the real world, thousands of different chemicals are present in our foods, and additional ones are generated when we cook the food. A few of these chemicals are potentially harmful—either being natural components of certain edible plants or found as contaminants, e.g., in seafood. Other chemicals are being added to foods to maintain or improve their freshness and extend their shelf life, or enhance their flavor and appearance.

What about the safety of such food additives?

DOI: 10.1201/9781003346661-13

Because these chemicals are deliberately added to foods, we could simply ban them for consumption and avoid exposure if they were not safe. Many food additives are non-hazardous to health under normal conditions, and therefore considered *GRAS* ("Generally Recognized As Safe") substances. They are not regulated and do not need to be approved. On the other hand, non-GRAS food additives need approval from regulatory agencies (keep reading for more on the safety assessment of chemicals).

The situation is very different for *unavoidable* chemicals in our food (residues, contaminants, or certain foods with a health hazard) that could potentially be toxic. For these latter cases, the regulatory authorities have set tolerance limits and defined maximally allowable levels in foods.

An off-the-cuff example is methyl mercury, which is present in certain seafood at various concentrations (for more on mercury see Chapter 13). Methyl mercury is an organic form of mercury formed by bacteria from metallic mercury in water or released from industrial waste. Unfortunately, methyl mercury accumulates in the natural food chain, causing biomagnification, as discussed earlier, with the highest levels found in predatory fish like tuna, swordfish, or marlin. Therefore, recommendations have been issued with regard to the amount of fish consumption that is still considered reasonably safe, considering even hearty eaters.

It is relatively easy to estimate how much of a hazardous chemical we ingest with our foods. The analytical labs routinely determine the concentrations of that chemical in different foods; those numbers are then multiplied by the estimated average daily intake of that particular food, and we can come up with what's called the *estimated daily intake (EDI)*. So far, we would know the daily, weekly, or yearly exposure—but how about the actual risk?

To gauge that risk, we should know, of course, how much is considered safe, or "acceptable," and how much would be past it. But here's the question that probably has been preying on your mind: how do we know what is an "acceptable" daily intake? In a nutshell, and jumping in ahead of Chapter 16 where we will discuss this issue in detail: we do have an estimate of the gray zone that separates "safe" from "no-longer-safe." For many chemicals, we do, in fact, have enough data about dose-toxic response relationships, obtained mainly from

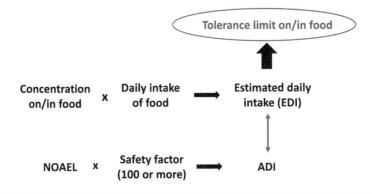

Figure 11.1 Calculation of tolerance limits of chemicals in food.

animals, which gives us a quantitative rating, the so-called no-observed-adverse-effect level (NOAEL)—stay tuned for more. We may also have data from epidemiological studies in humans, so that we can come up with a ballpark number that discerns what intake level of a chemical would still be considered safe; this is the so-called *acceptable daily intake (ADI)*. Now we can do the comparison; as long as the EDI (actual intake) is smaller than the ADI (the upper limit), we are fine, and the regulatory authorities can set the tolerance limits for that particular chemical in foods (Figure 11.1).

In addition to all those detailed calculations, one should also sit back for a moment and have a look at the big picture. Let's use common sense. For example, there have been sporadic reports in the media about certain spices (e.g., oregano, thyme, basil) that are contaminated with heavy metals, like cadmium, lead, or arsenic (more on these metals in Chapter 13). According to those reports, certain dried spices we buy in the supermarket are grown overseas, often in the presence of contaminated water that delivers the heavy metals to the soil where they are taken up by the plants. Cause for alarm? Are we slowly poisoning ourselves with lead, like the ancient Romans? Before falling into a depression, let's take a deep breath and look at the quantitative side (again) of the alleged threat. Frankly, what percentage of the amount of your daily food intake can be attributed to the few dried thyme leaves you're going to sprinkle on your fish dish? Analytical techniques no doubt allow for detecting traces of these metals in dried spices, but how much of the cadmium you unknowingly ingest each day can be chalked up to the skosh

Jill, cardiologist, and Joe, toxicologist, both passionate foodies, enjoying their night out

Figure 11.2   Different ways of reading a menu.

of oregano on your pasta? I bet it is a small, negligible fraction of the total exposure to those metals. That doesn't mean we shouldn't monitor contaminant levels in foods, including spices (that's a good thing), but we shouldn't blow up a risk that is low, and panic about it (Figure 11.2).

The problem is, we don't always know what exactly is in our foods. Sometimes we may be in for some big surprises, as illustrated with the next example.

Toxic Fries?

For starters, a bit of backstory. Some years ago, there was quite a turmoil about a chemical called acrylamide. *Acrylamide* is an industrial compound that has been widely used to make polymers (polyacrylamide). Its neurotoxicity in rats has been well documented, and neurological effects in workers in the polymer industry have been recognized. In addition, acrylamide is a rat carcinogen. For that reason, the chemical has been red-flagged, and safety limits have been established for occupational exposure.

For industrial chemicals, the US EPA sets such safe limits, termed *Reference Dose (RfD)*, which basically is the equivalent of an ADI value used for chemicals in foods, as outlined above. RfD is defined as the amount of daily exposure over a lifetime that can be predicted to produce no detectable health effects in humans. How exactly an RfD can be calculated is quite intricate and will be explored in Chapter 16, but so much here: in 2010, an RfD value for acrylamide based on neurotoxicity was calculated to be 0.002 mg/kg body weight/day.

What a shock when back in 2002 a research team around Margareta Törnqvist at Stockholm University, Sweden, reported that they'd found acrylamide to be present in some of our foodstuffs—and not just in traces, but up to several mg/kg food (especially in potato chips, bread, coffee, and other foods). Are you kidding? A neurotoxic compound and potential carcinogen in foods that we all consume daily, like bread, cereal, and potato dishes, and we didn't know about it? The first thing you do in science when something is reported that's hard to believe is to repeat the study. Many other researchers did so, only to confirm what the original study had found.

Nobody had been aware of that before. The alarm bells went off. The Reference Dose (or acceptable daily intake) for acrylamide was exceeded, not just by a few percent, but by more than ten-fold! What was going on?

Soon it became clear that acrylamide is not present in the foods from the very beginning, but that it is generated when the food is cooked at high temperature, e.g., fried or baked, not just boiled. (For nutritional chemistry freaks: the reason is that at high temperatures, sugars, and a specific amino acid—asparagine—react together to form acrylamide during the "browning" process. Potatoes are especially high in asparagine.)

To make matters worse, acrylamide has been classified by the IARC as a "probable human carcinogen." While with the estimated average daily intake of acrylamide we might be borderline safe with respect to the nervous system, the situation for cancer is different. Recall from Chapter 7 that theoretically there is no established "safe" dietary limit for genotoxic carcinogens, although this is now highly debated. How can this problem be cracked? Are we being exposed to something harmful that we cannot avoid? Should we quit eating potatoes and bread?

After the initial shock, the dust has settled quite a bit. After all, humans have cooked foods for thousands of years—although that is no irrefutable proof that acrylamide is safe. But numerous studies have refocused on acrylamide, and it became clear that the actual risk of developing cancer from acrylamide is smaller than initially feared. Here are some points to consider, that hopefully will lower your heart rate a bit:

First, it is not clear whether acrylamide, being a rat carcinogen, is genotoxic, i.e., damages the DNA; new evidence suggests that it may act in a non-genotoxic mode. If so, there would be a threshold in the dose-response curve, and an "acceptable" daily dose could be defined for humans. More recent studies have estimated an average EDI (estimated daily intake) for acrylamide of 0.0015 mg/kg body weight/day, or lower, which is not that different from the RfD (0.002 mg/kg body weight/day), established for neurotoxicity.

Second, we know that to become toxic, acrylamide must first be bioactivated, and metabolized to a form that is more toxic than the parent. This is what happens in the body; however, the DNA is not invariably damaged each time this happens. As discussed in Chapter 8, there are several defense mechanisms in place; one of them is glutathione, the ubiquitous scavenger of reactive, toxic metabolites. Glutathione readily captures the acrylamide metabolite and inactivates it. Better yet, the bioactivation step is slow, whereas the subsequent inactivation by glutathione occurs fast. So, the result of this complementary action is a net removal of most of the potentially dangerous chemical.

Third, and surprisingly, it was found that our body itself generates acrylamide, in fact, similar amounts as those ingested with food. Most probably, acrylamide is produced by the normal bacteria residing in our gut, the so-called microbiota (see Chapter 6).

Finally, it has never really been established by epidemiological studies whether acrylamide is a relevant human carcinogen. Studies from workers in the polymer industry, who are exposed to higher levels of acrylamide than others, do not have a higher incidence of cancer than control subjects who do not work in this industry.

The acrylamide saga is just another example of how careful we must be before making hasty conclusions that give the average consumer the horrors. There's no way we can entirely avoid exposure to acrylamide, but this chemical does not cause cancer in every person who will indulge in savoring potato chips just because it is present in baked or fried everyday foods.

Just My Cup of Tea

Okay, we get it, some potentially hazardous chemicals are already present in our foods from the beginning or may be generated during the cooking process. But how about those numerous contaminants coming from outside? How on earth does that stuff end up in our food? To take one example, how do some weird chemicals find their way into a neatly packed tea bag?

You read in a popular consumer report that a bunch of hazardous compounds have been detected in various brands of tea from different producers and origins, allegedly including cancer-causing agents and liver-damaging chemicals. In dried tea leaves? Really?

Being an astute learner, you of course ask the pivotal "how much?" question right away. As usual, the report doesn't talk a whole lot about numbers.

Among the list of toxic contaminants found in those tea samples are—apart from the notorious pesticides—*polycyclic aromatic hydro-carbons*, aka *PAHs* (e.g., benzopyrene, that we already talked about in Chapter 6). Plus, some chemicals you've never heard of. And, finally, something you're familiar with: iron filings.

Before you dump all your tea bags into the trash and switch to coffee for good, let's take a closer look. First, at the quantitative aspect.

The only reliable number you're able to find in that report is the amount of PAHs detected in the tea: it's in the range of 10 microgram (µg)/kg of dried tea leaves. One tea bag contains about 2 g of tea, so that would translate to 20 nanogram (ng) PAHs per bag (= 0.00000002 g). That seems like, well, an incredibly small amount. What does it mean? For comparison, it is about 500 to 1000-fold less than you would find on an average charcoal-broiled steak. No reason to panic (unless you are afraid of charcoal-broiled steak, as some folks are).

The second point to consider is the puzzling question asked initially: how do PAHs wind up in our tea bags? Let's recall that benzopyrene-type chemicals are generated by the incomplete combustion of organic materials like wood or charcoal. So, if there's a fire burning somewhere in the area where the tea is processed, PAHs would inevitably be released into the air and you would be able to find tiny traces of those chemicals on the leaves (probably on other agricultural products as well)—unavoidable, and quantifiable with modern cutting-edge technology, but not *a priori* a human health problem. Moreover, PAHs are not water-soluble, so they might stay with the tea leaves and not in the tea.

Further down the list are some arcane chemicals: *pyrrolizidine alkaloids.* This is a group of naturally occurring compounds found in several hundreds of plant species. The chemicals at high exposure are toxic to livestock and, rarely, humans. There are isolated reports about high levels of pyrrolizidine alkaloids in certain herbal teas, honey, and spices, but normally the amounts are too small to pose a significant risk. But indeed, some of these unsolicited plants may end up as contaminating herbs during the harvest process of the tea leaves—the big question is, how much?

Average pyrrolizidine alkaloid levels in black tea, according to a European study, are some 40 µg/kg (again, assuming a tea bag contains 2 g of tea leaves, this would result in 0.08 µg pyrrolizidine alkaloids per serving). Recommendations by the European Medicines Agency (EMA) are 0.35 µg/day as a permitted limit for adults, with a safety factor, aka margin of exposure (MOE, to be explained in Chapter 16), of greater than 10,000. So, we should be safe. Fortunately, with increasing awareness over the years, the contamination levels of pyrrolizidine alkaloids in foods have been declining. (An exception is the sporadically reported life-threatening toxicity from pyrrolizidine alkaloids from herbal teas in areas where people have harvested plants containing these compounds.)

Traces of other contaminating chemicals may find their way into the tea bags from packaging materials. And the iron filings mentioned above likely stem from the tea leave-processing machines; they will not end up in your cup.

So, before blowing the "contaminated" tea issue out of proportion, factor in the exposure. Weighing the small risk against the health benefit of teas, the result clearly is in favor of the benefit (to be discussed in Chapter 18). You, together with zillions of other folks, may continue to enjoy your cuppa.

Invisible and Unavoidable: A Mold Toxin

If food goes moldy, throw it away. Our grandmothers told us that.

There is a truth in it; it's not only the repugnant moldy smell but foods can also be infected with certain molds that produce toxic chemicals. Some of them are highly toxic, and one group among them belongs to the most potent chemical carcinogens known.

Under conditions of high temperatures and humidity, a mold called *Aspergillus* flavus can infect a number of plants or their stored food products (wheat, corn, rice, chili peppers, peanuts, pistachios, sesame seeds, and others) and produce a toxin called *aflatoxin* (abridged from the mold's name). Aflatoxin, strictly speaking, a family of several related aflatoxins, is acutely toxic, but what is more worrisome is the fact that it can cause liver tumors in people. This has been especially problematic in certain parts of China and Africa. Interestingly, but explicable, organic crops are more susceptible to contamination with the aflatoxin-producing mold than fungicide-treated crops.

Is aflatoxin a real problem, or just another hype making the headlines?

Make no mistake; unfortunately, this time we're talking about a significant issue.

Rats develop liver cancer when exposed to aflatoxin. Mice, in contrast, are highly resistant—the question is, are humans more like rats or mice? I'm afraid on that score it's the former.

Aflatoxin is converted in the liver to a reactive metabolite that avidly binds to DNA and induces mutations and eventually cancer (see Chapter 7). Mice, unlike rats, have a very efficient detoxication system involving glutathione that traps the reactive metabolite before it can interact with DNA. Humans are not that efficient. Aflatoxin has been classified by the IARC as a human carcinogen because the evidence has been overwhelming.

In view of these risks, foods are being randomly sampled by the appropriate governmental agencies to monitor possible contamination with aflatoxin. Highly contaminated foods that exceed the acceptable limit are pulled from distribution. For example, the US FDA has set an action level of 20 ppb for aflatoxin. What does that mean in lay-people's terms?

First, an *action level* is a level for enforcement by the regulatory authorities and is in place for *inadvertent* residues, like aflatoxin produced by mold—obviously beyond our control. In contrast, a *tolerance level* is a level of a contaminant as a direct result of a *purposely applied* chemical, e.g., a pesticide.

Second, when reading reports on residues in foods or drinking water, you will often encounter the terms, *ppm, ppb, ppt*. These abbreviations stand for "parts per million," "parts per billion," and "parts per trillion," respectively; ppm is an alternative way of saying milligrams per kilogram (one kilogram has one million milligrams, etc.). In our example, the action level (20 ppb) means, no more than 20 µg aflatoxin per kg of food is allowed or else that batch must be pulled from the market.

Obviously, it will not be possible to eliminate aflatoxin and similar mold toxins from our food altogether, especially in tropical or subtropical regions. And nobody can be blamed that the toxin is there.

All we can do is try our best to minimize the exposure.

Trans Fat

Trans fats have fallen into disrepute because of their potential to increase the risk of heart attacks. (The name "trans" stems from their chemical configuration that is slightly different from the other unsaturated fats.)

Trans fats find their way into the diet via two sources. First, they can be formed by bacteria in cows and other ruminants and they end up in milk and butter. Second, they arise during the industrial processing of liquid vegetable oil, resulting in "hardened" fats such as margarine or spreads. High intake of trans fats has been associated with changes in blood cholesterol levels—increases in the "bad" cholesterol (LDL) and decreases in the "good" cholesterol (HDL).

So, what is the tolerable upper limit?

Our standard calculations for tolerance levels don't work well here because we have no clear idea of how much trans fats increase the risk of cardiovascular disease significantly. The fact is, several countries have banned the industrial use of trans fats (the United States in 2018) or regulated it (European Union in 2021, to less than 2 g of trans fat per 100 g fat). But despite the bad reputation of trans fats, we shouldn't forget that they are just one of the multiple factors involved in an increased risk for cardiovascular disease (reverting to Figure 9.2). Nevertheless, avoiding them won't hurt you.

TAKE-HOME MESSAGE

- Foods contain many potentially harmful natural components; the exposure determines whether or not they are "safe."
- Some food additives are GRAS (generally recognized as safe); no approval is needed. Potentially harmful additives are avoidable and can be banned. For others, tolerance limits (ADI) can be set.
- Acrylamide is generated spontaneously in foods (potatoes, coffee, bread) cooked at high temperatures. It is a rat carcinogen, but the risk of developing cancer from food-derived acrylamide in humans is low.
- The Reference Dose (RfD) is the amount of daily exposure to a chemical in the environment that can be predicted to produce no detectable health effects in humans.
- Benzopyrene and other PAHs might end up in your food (or teas) via the air—they can be detected, but the trace amounts are too low to be a significant human health risk.
- Aflatoxin, a potent human carcinogen, is a chemical produced by molds. Small amounts are found on certain foods; because one cannot avoid the contamination, action levels are set.
- *Special*: The most dangerous factor in eating too many cookies is not cancer-causing chemicals but likely the extra calories.

12

DIETARY SUPPLEMENTS

The More the Better?

Boosting Health with Vitamin Supplements?

The sobering news first, even though it might come as a mild shock to you: in healthy, well-nourished people, vitamin supplements have little (if any) proven beneficial effects for human health *unless* there's a vitamin deficiency in the first place.

"But vitamins are good for you," you might argue, "so it can't hurt to pop a pill each day?" Wrong again. Not only are excess vitamins (the bulk of the water-soluble ones) excreted quickly in the urine, turning the health-promoting exercise into a futile activity, but overdosing on some (especially the fat-soluble) vitamins can also increase the risk for adverse effects.

Per definition, vitamins are required by the body in small amounts, which one normally achieves by eating a balanced diet. Excess vitamins, taken as supplements, won't make you super-healthy.

We often tend to forget that the "exposure concept" (how much, how often, and by what route), discussed throughout this book at length (some might say, relentlessly hammered into the head of the patient reader), is also applicable for "wholesome" chemicals that we take deliberately. There is a clear dose-response relationship for the good stuff too. And that pertains to vitamins and other dietary supplements as well.

Dietary supplements are generally taken on top of the regular diet, with the goal to provide additional health benefits. They are thought to enhance protection against cardiovascular disease, cancer, premature cognitive decline during aging, or other diseases. Despite the wide use of dietary supplements, their benefits are often questionable (mind you—in contrast to pharmaceutical drugs, the manufacturers don't have to provide evidence for efficacy, and clinical trials are not required before such supplements are put on the market). While they are generally safe when taken in moderation, the risk for adverse effects increases with increasing doses. For example, excessive doses of vitamin E have

DOI: 10.1201/9781003346661-14

resulted in bleeding, weakness, and even increases in the recurrence of tumors in cancer patients. Similarly, it was shown in controlled clinical studies that high doses of vitamin A (or its precursors) were associated with an increased risk of lung cancer and prostate cancer, as well as toxicity to bones. In addition, congenital abnormalities have appeared in babies born to women who had consumed large doses of vitamin A during pregnancy.

A similar situation applies to the so-called *nutraceuticals*, i.e., purified products derived from human foods (mostly fruits and vegetables). They are widely believed to provide additional health benefits beyond the basic nutritional value of those foods. That makes sense as many nutraceuticals are chemicals with clear antioxidant effects (see Chapter 8) or anti-inflammatory properties—so they should, theoretically, protect against those chronic diseases where oxidative stress and inflammatory processes are involved, right? But—not so fast. In cells in a culture dish, these compounds are often very effective. In a live organism, that's quite another story. As we'll see in the following section, once again the dose is the key—and many gimmicky stories about magic antioxidants have a surprising twist in the tail.

Green Tea, Red Wine, and Dark Chocolate

Many foods rich in the natural substances, *polyphenols*, are purportedly good for you because they are associated with a lower risk of developing certain diseases including cardiovascular disease, cancer, or cognitive deficiencies. Especially famous are the flavonoids in berries, flavonols in dark chocolate, catechols in green tea, or resveratrol in red grapes. We know that polyphenols are potent antioxidants. But is this knowledge enough to make a causal association between the compounds' antioxidant properties and their beneficial effects? Let's have a closer look.

First of all, it is important to realize that most findings with polyphenols stem from *in vitro* research, and there is no evidence that the same happens in live humans as well.

The second point to consider is the polyphenols' limited bioavailability. Although polyphenols are readily absorbed from the gut, they are rapidly metabolized, and the metabolites don't have antioxidant properties. The remaining concentrations that actually get into the body are low, in fact too low to exert a powerful antioxidant effect.

One of the most famous polyphenol nutraceuticals is *resveratrol*, a naturally occurring substance abundant in grapes and red wine. *In vitro*, resveratrol clearly is an antioxidant at relatively high concentrations. The putative protective effect of resveratrol in wine has even been invoked to account for the so-called "French paradox," i.e., the observation that there is a relatively low incidence of cardiovascular disease in the French population despite their heavy consumption of saturated fats and meat. So, is resveratrol a panacea? Before you reach for the corkscrew and sacrifice another bottle of your favorite claret, hang on a second. Resveratrol concentrations in red wine are relatively small, and trying to increase the dose by maximizing the number of refills is not worth the headache the next morning. Also, the bioavailability of resveratrol is extremely poor, and its metabolites are rapidly excreted. So, is the only solution to switch to high-dose, purified resveratrol as a commercial nutraceutical? It probably won't harm you, but there are no convincing clinical data available yet to demonstrate that resveratrol is beneficial for human health *in vivo*. While *in vitro* data and animal studies show clear beneficial effects of resveratrol, the clinical trials so far have not consistently demonstrated beneficial effects. As always, caution is warranted as certain adverse effects may emerge at exceedingly high doses.

Before concluding that the hype about polyphenol antioxidants is a myth, think about the well-known beneficial effects of eating fruit and berries and drinking tea or coffee (let alone the soothing effect of dark chocolate). Foods contain a host of other beneficial substances that might work in concert. Don't expect an isolated chemical derived from food and taken as a high-dose supplement to have the same beneficial effect as the whole food that contains the same and other beneficial chemicals. Too much of one thing might turn on an adverse response—sounds trivial, so climbing higher up the dose-response curve should always be carefully evaluated.

Herbal Supplements

Herbal medicine has been used for thousands of years, and herbal dietary supplements are more popular than ever. Everybody is familiar with ginseng, echinacea, ginkgo, turmeric, or the myriad of different herbal teas. When used in wise amounts that's fine; the opinions are divided, though, when it comes to their efficacy.

However, contrary to popular belief, the fact that herbal supplements are of natural origin (botanicals) does not guarantee that they are safe (you may want to refer to Chapter 5). I'm not talking about the rather shocking findings that some of them are adulterated ("spiked") with pharmaceutical drugs. Problems may arise when the doses are increased beyond the recommended dosage. In many cases, not the whole plants or parts of them (e.g., leaves, flowers, bark, or roots) are marketed, but highly concentrated extracts from these plants. For example, green tea is being enjoyed by millions of people, and there is no doubt that a cup of green tea is beneficial for you. However, there are numerous reports that green tea extracts, taken at excessive dosage, may cause toxicity to the liver. Toxicity has also been documented for other herbals, and the target organs, besides the liver, include the cardiovascular, nervous, and digestive systems as well.

Another tricky thing that is easily overlooked is that in the United States herbal supplements are regulated in a different way than pharmaceuticals, although they allegedly have a certain therapeutic benefit. As long as they are promoted by the manufacturers as "health-enhancing" or "improving the general well-being" or carry a similar attribute, and not referred to as treating a certain disease, the company that manufactures and sells the product is not obliged to provide evidence that the product is safe. Instead, should there be incidences of adverse reactions or plain toxicity caused by these supplements, the regulatory agencies must demonstrate that those products cause harm—and they don't always have the time and resources to do so. Ironically, the situation is quite the opposite for drugs, pesticides, food additives, etc., where the "burden of proof" lies with the company. The demonstration of efficacy, toxicokinetics, of a dose-toxic response relationship and its underlying mechanisms, in animals and then in humans entails enormous costs and takes time (for much more on this keep reading, Chapter 15), but as a result, we have a pretty good idea about the hazard, exposure, and the human health risk.

Unfortunately, this is not the case for dietary supplements including herbals. The potential human health risk of herbal dietary supplements has been incompletely studied, and little information is available on the quantitative aspect, i.e., the dose-response relationship. What holds true for all other chemicals is also real for the natural, allegedly wholesome herbal supplements: the exposure, dose, and dosage determine at what point the beneficial effects will no longer outweigh the increased health risk.

TAKE-HOME MESSAGE

- Dietary supplements and nutraceuticals are not regulated by the health authorities; they don't need to be approved in order to be marketed, and evidence to show efficacy is not required.
- The dose-response concept also holds true for "good" chemicals (e.g., vitamins and other dietary supplements). Too much of a good thing might become toxic.
- It is not known at this time, even unlikely, that polyphenols, the naturally occurring antioxidants in green tea, red wine, berries, and dark chocolate, are antioxidants *in vivo* at doses that humans can reasonably consume. They get rapidly metabolized, and their bioavailability is low.
- Herbal dietary supplements are generally safe at reasonable dosage but might become toxic at excessively high concentrations (e.g., as extracts) and exaggerated intake.
- *For dessert*: If you take something that at a low dose doesn't work and at an excessive dosage might even be harmful—think twice about whether you really need it.

13
SIGNIFICANT CHEMICAL RISKS
Persistent and Widespread

Chemicals That Don't Make the Headlines (Anymore)

Many of the examples we have discussed so far concern risks from toxic chemicals that many people perceive as highly significant, if not alarming. Dangers lurking in our foods, in the air, the water, in everyday consumer items, and everybody seems to be talking about them. We have seen that oftentimes the reality contrasts with those media-driven perceived risks. Upon closer look, when the actual exposure data are factored in, the human health risk for some of those compounds turns out to be relatively small, if not negligible.

In this chapter, we will consider the other side. There are real risks from toxic chemicals that hardly anybody is talking about, although they are of great concern worldwide. The headlines about them are less flashy, the articles less sensational, and once they flare up, they quickly subside like a stifling flame. However, the problems are real.

Yet, it is old business. We've known these challenges for a long time, but somehow, many of us have put them on the back burner. Maybe it is because we have become jaded to them after a while, or because many of the problems occur "elsewhere" and not directly in our backyards, or because we are simply not aware of the scope of the issue. I'm talking about certain metals and metalloids in our environment (lead, cadmium, mercury, arsenic), some volatile compounds that are generated during the burning of organic matter (benzene, dioxin, and related compounds), wood smoke contributing to air pollution by the release of ultra-fine particulate matter, and hazardous pesticides.

The examples we'll be discussing below are among the top ten on the list established by the WHO as *chemicals of major public health concern.*

DOI: 10.1201/9781003346661-15

Arsenic

Arsenic is a naturally occurring element (a so-called metalloid, i.e., chemically standing between metals and non-metals) present in the ground, but also a toxic by-product arising from industrial use. The major route of exposure to arsenic is ingestion from drinking contaminated water. Arsenic toxicity has been recognized to be a major problem in certain areas (e.g., Bangladesh, Taiwan, or Chile) where deep artesian wells have been used to tap the water, rather than shallow wells or surface water, which are often contaminated and a cause for bacterial infections. The flipside is that deeper layers can contain relatively high levels of arsenic.

If you are a classic-mystery aficionado, you may associate arsenic with acute poisoning (arsenic has often been called the "poison of kings" and the "king of poisons"). More significant, however, are its chronic effects. Arsenic can be toxic to the nervous system; it has also been associated with cardiovascular toxicity and diabetes. Importantly, though, arsenic can cause skin, bladder, and lung cancer. The evidence is so compelling that arsenic has been classified as a known human carcinogen. The WHO estimates that more than 40,000 people die annually in Bangladesh alone because of high exposure to arsenic.

As enough reliable human data have become available, a "safe" exposure level (threshold) for skin cancer has been set, below which no toxicity is to be expected. (For more on how to exactly calculate such tolerance limits keep reading; Chapter 16.)

You may raise a legitimate objection here—didn't we say earlier (Chapter 7) that setting a dose threshold for genotoxic cancer-causing chemicals is a subject of debate because we cannot exclude that even the smallest doses can be harmful? A dose threshold would mean that mild exposure to arsenic would be tolerated by the body, with no apparent toxic response. Having worked your way up to this point of the book, you might have a ready explanation for the basis of this tolerance phenomenon: our body has defense mechanisms in place to protect against toxic chemicals, and an important one is "detoxication" by metabolic conversion. That's exactly the case here.

In fact, arsenic is converted in the body to different metabolites, some of which (but not all) are much less toxic than the parent arsenic.

So, at low concentrations, arsenic is being inactivated (metabolized and excreted). However, at higher concentrations, the cells can no longer compensate for the reaction of activated arsenic with various proteins due to the increasing body load of the parent arsenic; the system becomes saturated, and toxicity might ensue. The critical threshold of concern is an approximate concentration of arsenic in the water of 0.25 mg/L. Unfortunately, in many countries, millions of people are being exposed to concentrations of arsenic in their drinking water that exceed that limit.

Major public health programs are aimed to screen water supplies (the maximum contaminant level set by the US EPA and recommended by the WHO has been set at 0.010 mg/L). Also, worldwide efforts are underway to support new arsenic-removal technologies.

Mercury

Mercury is a naturally occurring metal that exists in different forms (as an element or as more complex compounds), each causing different types of toxicity. It is released into the environment by mining processes and waste incineration, or by burning coal. Exposure to mercury occurs mainly by inhalation and via consumption of contaminated fish and shellfish. Metallic mercury is highly toxic after inhalation of its vapors, which primarily damage the kidneys, but is not well absorbed in the gastrointestinal tract when ingested.

Toxicity from mercury was first observed in the 1950s in western Japan, in a region around Minamata Bay, where mercury was released from an industrial plant into the sea. Aquatic microorganisms converted mercury into the dangerous methyl mercury, which in turn kept accumulating in fish and shellfish. In populations who consumed a lot of contaminated fish, neurotoxic symptoms occurred, starting with abnormal sensibility, and disturbed coordination, then progressing to difficulties in speech and vision. The reason remained a mystery for a long time because mercury had not been known to have toxic effects on the brain. Only in the 1960s, the culprit, methyl mercury, was identified for what became known as the Minamata disease. Chronic ingestion of methyl mercury can be harmful to the unborn child and can impair cognitive function in children.

Harking back to Chapter 8, we talked about our own body's defense mechanisms that could shield off certain toxic effects after a mild chemical insult. One of those protective compounds, glutathione (and related species) comes into play when defending against mercury or methyl mercury. The trapped and thus "inactivated" metals are normally excreted via the kidneys, but something unexpected can mess up that process—let's have a brief look at this slightly off-topic curiosity.

After binding to mercury, a part of the glutathione molecule is cleaved off as a normal process in our body. But this truncated, smaller molecule plays a trick on the brain, fooling the normally selective access control (the molecular doorkeeper): because the foreign molecule resembles a normal, physiological amino acid, methionine, both in shape and characteristics, the blood-brain barrier cannot distinguish the two from each other and grants methyl mercury (emulating methionine) access. Once in the brain, methyl mercury can unfold its toxic effects, like the warriors hidden in the Trojan horse. In fact, even small concentrations of methyl mercury can kill brain cells, especially in those areas that are responsible for coordination.

So, for us consumers who occasionally enjoy a fish dish—how safe is it to consume seafood? You guess the answer—it's all about exposure, rather than an absolute yes/no answer. International agencies have issued guidelines for consumers, balancing the risk against the beneficial (nutritional) effects of eating fish. For an average consumption of fish, the amounts of methyl mercury are mostly below the tolerated levels; for frequent fish consumers, and especially for pregnant women or parents with small children, a choice should be made and consumption of high mercury-containing fish species (e.g., swordfish) should perhaps be limited.

Currently, great efforts are underway to reduce the industrial production of mercury and promote safer use.

Cadmium

Cadmium is another metal that is toxic to humans. The levels in the natural environment are low, but through its industrial use (e.g., for electronics and batteries) the potential exposure to cadmium has increased. Humans have been exposed to cadmium from food (shellfish, shrimp) or inhalation (tobacco smoke or occupational exposure). Accumulation in the kidneys can lead to kidney toxicity; cadmium has also been classified as a human carcinogen.

Similar to the story about methyl mercury (above), an endemic disaster has been the wake-up call to recognize the toxic potential of cadmium. In the 1960s, cadmium-containing mining wastes in certain areas of Japan were dumped into rice paddies from where the metal got into the water and then absorbed by the plants. Sometime later, it was noticed that women in the area developed severe pain in their bones, back, and shoulders—in fact, the then-unknown disease was called Itai-Itai disease (meaning "ouch"). It turned out that those exposed had a daily intake of cadmium that exceeded the typical intake in people living elsewhere by several hundred times.

Nowadays, tolerable daily or weekly intake levels for cadmium have been established, and the use of cadmium has been restricted.

Again, it seems interesting to have a quick side glance at how cadmium exerts its toxicity in the kidneys. Like the other toxic metals, cadmium produces oxidant stress (need a brush-up on oxidant stress? Turn back to Chapter 8). And like other toxic metals, cadmium can be trapped by glutathione and related compounds in our body. But the body's defense goes even further: upon "sensing" cadmium, the liver produces a specific protective protein called metallothionein that can capture cadmium and thereby prevent it from doing any harm. The liver releases the bound and inactivated cadmium into the bloodstream, from where it reaches the kidneys. However, instead of being flushed out, the metallothionein-cadmium complex is taken up by the kidney cells. While the cells eventually degrade the metallothionein protein, the previously trapped cadmium is freed and accumulates in those cells until the levels are high enough, so that damage to the kidney ensues.

Currently, international intervention programs aim at minimizing cadmium emissions from mining sites and waste management plants, recycling cadmium-containing products (in electronics), and eliminating cadmium from consumer products. However, because once absorbed cadmium stays in the body for a long time (decades rather than years), dealing with cadmium toxicity has been a challenging project.

Lead

Lead is a naturally occurring toxic metal in the Earth's crust and has been used for thousands of years. Its wide industrial application has entailed extensive environmental contamination and caused widespread human health problems.

Exposure to lead occurs primarily by inhalation and ingestion of dust particles. Lead has no known biological role in the body; instead, the metal can cause oxidant stress and disrupt what's called "cell signaling" (how cells talk to each other). Excessive lead is stored in bones and teeth and is thus inactivated, but free lead can reach other target organs. The most vulnerable ones are the brain and bone marrow (where blood cells are formed), followed by the gastrointestinal and cardiovascular systems, as well as the kidneys. Children are particularly sensitive to lead, and impairment of cognitive function is a worldwide problem.

Finally, lead, like most other metals, is carcinogenic. It causes tumors in the kidneys, lungs, and brain of laboratory animals, and epidemiological studies have solidified a link between lead and cancer in lead-exposed workers. As the evidence for a causal link between lead and tumors kept increasing, the IARC reclassified (in 2006) lead as a "probable human carcinogen."

Efforts are underway to reduce the non-essential use of lead. Although in many countries exposure to lead-containing gasoline, paint, or plumbing has been greatly reduced, resulting in a clearly lowered body burden of lead in the general population, lead toxicity is still a major problem in developing countries.

Benzene

We have already come across benzene in Chapter 8 and have learned that benzene can produce oxidant stress, especially in the bone marrow, which is poorly protected against damaging effects.

Benzene is a highly volatile chemical generated when burning organic matter and petroleum-based materials. Exposure of humans to benzene occurs primarily by inhalation from solvents, tobacco smoke, motor fuels, and occupationally. The chemical is highly toxic to the blood cell-forming organ (bone marrow) and has been classified as a human carcinogen.

Depending on the exposure level, benzene can cause anemia (a depletion of red cells and white blood cells that play a huge role in immune defense). An investigative study in shoe factories in China revealed that workplace-exposed people exhibited greatly reduced numbers of different types of blood cells, even at values that were below the currently used occupational exposure limits. Because all blood cells were affected (and not just, say, the red cells) the most logical explanation was that a toxicant likely damaged the bone marrow. It is also known that at higher and longer exposure, leukemia (blood cell cancer) can occur.

To take remedial action, international programs have been aiming at replacing benzene from solvents and reducing its use in industrial processes.

Wood Smoke and Air Pollution

Wood smoke is often considered harmless and benign as it is something "natural" (didn't we encounter this widely held opinion before?). However, wood fires (from stoves, fireplaces, or wildfires) emit health-damaging chemicals that contribute to air pollution and pose a serious human health concern.

In the United States, wood is being used as a primary or secondary method of heating in about 12 million homes. Although that number is impressive, the developed countries are not the major problem. Let's recall that worldwide some three billion people (that's roughly half of the world's population) depend on solid fuels for their daily heating and cooking—often indoors, without chimneys or other proper air ventilation systems. Indoor air pollution in households from such fires is among the leading risk factors for death or morbidity worldwide. In fact, according to estimates by the WHO, one to two million people globally die prematurely each year as a direct consequence of indoor air pollution by smoke.

Wood consists of cellulose and lignin, but the combustion process breaks these structures down into smaller chemicals. As a result, a host of different volatile chemicals and particles are generated. In fact, wood smoke contains thousands of different chemicals; some of them, e.g., benzene or dioxins, have been classified as carcinogens. (Among the better-known gases generated by burning solid fuels are carbon monoxide, CO_2, ozone, and nitric oxides that contribute to local, regional, and even global air pollution and even climate change.)

Apart from the volatile chemicals, the *particulate matter* emitted by these fires poses an equally serious challenge. Especially dangerous are the fine particles called PM2.5 (particulate matter, smaller than 2.5 μm—i.e., not visible by the naked eye, and about 50 times smaller than the diameter of a human hair). The problem is that PM2.5 can reach the deepest parts of the lung, where they are deposited and can damage the tissue. Another problem related to their small size is their persistence in the air; if they are generated outdoors, they can easily penetrate houses. They can stay in the air for a long time and become air-freighted over long distances. Both the volatile chemicals and the fine particles can cause oxidant stress and inflammation in the body; they can also cause allergic reactions.

There are hardly any controlled studies in humans on the adverse effects of wood smoke, but those available suggest the potential for inflammation and cardiovascular effects. On the other hand, there are many epidemiological studies, i.e., studies that look at patterns and the distribution of health effects, related to wood smoke. There is strong evidence for respiratory problems and pneumonia in children, COPD (chronic obstructive pulmonary disease) in adults, and the development of cataracts leading to blindness. (Lung cancer, although a concern, seems to be more related to coal combustion than wood burning.)

Most of the toxicity studies in lab animals confirm what we've learned from the human data: short-term, high-dose exposure to wood smoke weakens the immune defense of the lungs, causes inflammation, and can lead to long-term or even permanent damage to the lungs.

Our joint efforts must go toward reducing the emissions from burning solid fuels, in particular burning wood, farming refuse, or dung, all of which entail large emissions of pollutants in relation to the low levels of energy gained. Air pollution (especially indoors) is a significant environmental health risk factor.

Dioxins

We are talking here about a whole family of related compounds (dioxins) including "dioxin-like" chemicals, denoted by complicated chemical lingo. The most toxic dioxin among them is TCDD (2,3,7,8-tetrachlorodibenzodioxin, but to keep things simple we will refer to TCDD by its popular albeit sloppy name, "dioxin").

Dioxin is inevitably generated by several chemical reactions during certain industrial procedures, but also by combustion processes, e.g., in waste disposal power plants or forest fires. It is released into the air, although in low quantities. The real problem with dioxin and its related compounds is that those chemicals are persistent in the environment. Also, they can bioaccumulate in organisms because of their long biological half-life, in fact, they stay there for years.

Dioxin (TCDD) is a human carcinogen, likely acting by non-genotoxic mechanisms (not directly damaging DNA). Furthermore, in rodents, dioxin can cause malformations in the embryo and has been shown to be toxic to the immune system. Its varying degree of acute toxicity across different animal species is striking; guinea pigs are extremely sensitive (0.0005 mg/kg body weight is lethal), whereas hamsters are at the other end of the sensitivity scale (they are ten thousand times less sensitive) for reasons unclear to date. What do we know about dioxin's hazard in humans?

Unfortunately, we had to learn a bitter lesson through another industry accident. In 1976, in Seveso, northern Italy, dioxin escaped from a reactor used to produce a chemical required for the synthesis of an antiseptic. Several thousand acres of a densely populated area were contaminated by the incident, dumping several kilograms of dioxin into the environment. People were evacuated (fortunately, nobody lost their life), but thousands of animals died. The major health effects in humans were skin lesions in children (called "chloracne"). In the past, dioxins were also present as a contaminant of Agent Orange, which was used in the Vietnam War. Although causal relationships are always difficult to establish, the available evidence suggests that there is an increased risk for developmental effects in children exposed to dioxin.

Why is it so difficult to find out how exactly dioxin (and its related chemicals) affects human health (and animal health too)? What is known is that dioxin avidly binds to a specific receptor in the cell, called AHR. Upon binding, the receptor gets activated and migrates into the cell nucleus; once inside the nucleus, the dioxin-receptor complex binds to specific regions of the DNA where multiple genes are "switched on." The AHR's key role in mediating dioxin's toxicity was further solidified as we've learned that mice whose AHR had been genetically removed are highly resistant to dioxin. What we don't understand yet is how the interplay of all these activated genes together produces toxicity. We don't have a clear picture yet, but it has become evident that genes involved in cell death (apoptosis), oxidant stress, development, and fat metabolism play a role—if you are getting confused, be reminded that such an analysis is confusing for the researchers too, but to discuss the mechanisms in detail is beyond the scope of this book.

Dioxins are not being directly used in any industrial application; rather, they are an unavoidable product from combustion processes. Therefore, all we can do is to reduce emissions by developing and applying appropriate burning technologies. In addition, the levels of dioxin are carefully monitored in food, air, human milk, and at the workplace (occupational exposure).

Structurally related to dioxins are two other classes of chemicals, the dibenzofurans and the polychlorinated biphenyls (the latter are better known by the abbreviated name, PCBs). They share the same untoward properties, i.e., persistence in the environment and bioaccumulation. PCBs are no longer manufactured in the United States, and their use in industrial products is phasing out. The compounds, though, will be around for quite some time.

Hazardous Pesticides Revisited

Pesticides can be a potential risk because of both their acute and chronic toxicity. Serious health effects including fatalities have been reported, mainly due to occupational exposure and accidental (and sometimes, sadly, suicidal) poisoning. In addition, as elaborated in Chapter 10, environmental contamination with pesticides can indirectly lead to human exposure via food or water, although the exposure will be at much lower levels.

The reasons for concern are two-fold: the highly hazardous nature of some pesticides and the long half-life in organisms, and persistency in the environment of some others.

First, let's take a closer look at the broad range of acutely hazardous pesticides. Their potential hazard, e.g., from handling commercial products containing these agents, has been classified by the WHO. The characterization ranges from "extremely hazardous" (e.g., mercuric chloride, parathion), based on an oral LD_{50} in rats of less than

5 mg/kg body weight, to "highly hazardous" (e.g., arsenate, nicotine), to "moderately hazardous" (e.g., copper sulfate, DDT, paraquat, rotenone), to "slightly hazardous" (e.g., glyphosate, malathion, and other pesticides with an oral LD_{50} greater than 2,000 mg/kg body weight). Guidelines for the safer use, handling, and storage of highly hazardous pesticides have been established accordingly.

On the other hand, the issue with chronic toxicity of pesticides is their environmental burden due to their persistence, as well as the insufficiently characterized long-term effects for humans and the environment. Most of the old, persistent pesticides (e.g., DDT, dieldrin, mirex) have long been banned in the United States and Europe, but are, unfortunately still used in some developing countries. That doesn't mean that all the other pesticides are harmless. Degradation in the environment takes time. More importantly, and a bit worrisome, the hazard and risk of the numerous degradation products are not always known. For these persistent pesticides, exposure levels must be monitored on a global scale. In addition, the search for less environmentally burdening pesticides must go on (green chemistry, see Chapter 20). Importantly, though, the risks should also be balanced against the benefit of these compounds, putting us in a quandary.

Taken together, the chemical risks described in this chapter have long been recognized but are often underrated by many people. However, the threat is significant, posing a worldwide human health problem and an increasing environmental risk. Among the reasons are the ubiquitous nature of many of those chemicals, the unavoidable exposure, their longevity and persistence, the potential to bioaccumulate, and their carcinogenic potential in humans. The good news is that for most of them we have a pretty good idea about the dose-toxic response relationship. Therefore, efforts to keep the exposure levels below the thresholds of concern should have priority.

TAKE-HOME MESSAGE

- Metals like lead, mercury, and cadmium are ubiquitous in nature and found in various foods and manufactured products. We are constantly exposed to low levels of them—knowingly or unknowingly. Chronic toxicity is the problem because these metals can accumulate in tissues, produce oxidant stress, and can be carcinogenic in humans.

- Arsenic is a human carcinogen. Its major source of exposure is contaminated drinking water in certain parts of the world.

- Methyl mercury is neurotoxic and can cause cognitive deficits in the brain of unborn children. Its major source of exposure is contaminated food (seafood), whereas for metallic mercury it is inhalation.

- Cadmium is a human carcinogen and produces kidney toxicity due to accumulation in cells. Exposure is through food (shellfish) and inhalation (cigarette smoke, occupational).

- Lead is a probable human carcinogen (kidney) that can damage the brain and impair normal blood cell formation. Children are particularly sensitive.

- Benzene is a volatile chemical present in tobacco smoke, gasoline, and industrial solvents. Inhalation exposure can lead to bone marrow (blood cell) toxicity and leukemia.

- Wood smoke, especially from indoor cooking and heating, is a major cause of air pollution and a significant human health risk factor.

- Dioxin (TCDD) and similar compounds are generated by combustion processes. Despite the low concentrations in the air, the real issue is their persistence in the environment and bioaccumulation. Dioxin is a carcinogen and has a multitude of potential adverse effects.

- Pesticides pose a human health risk because of two characteristics—some are persistent in the environment and bioaccumulate in organisms, while others are highly hazardous.

- *By the way*: The most significant human health risks do not necessarily make the daily headlines.

14
DRUGS

Adverse Drug Reactions

A fictional (yet familiar) setting: you're at a summer night's party, pirooting around, a sweating bottle of locally brewed beer in one hand, some finger food in the other, engaged in small talk, trying to avoid heavy topics like politics or religion—but without warning, you're getting entangled in a discussion about pharmaceutical drugs. Uh-oh.

"Those pills are dangerous," a woman who has just joined the group interjects, swirling up her kale smoothie. "Toxic. I prefer natural remedies. Drugs can do bad things."

Meanwhile you should have gathered enough arguments to convince a chemophobic person that synthetic, human-made drugs are not always bad. In fact, they have a lot of features that come in handy: they are clean, they have been tested, and, importantly, they work. They have been developed to help the body fight specific diseases and to maintain or restore deficiencies in the body.

"Of course, they can do bad things, sometimes," you venture. "But it all depends on the dose."

"Here you go," Kale Smoothie says, throwing up her free hand. "Synthetic drugs. It's all about profit, profit, profit. At the expense of safety." She has obviously taken umbrage at our presumed siding with big pharma.

You check yourself halfway through an eye roll. "Pharmaceutical drugs are biologically highly active, they are *designed* to trigger something in your body—if they didn't, they would be lousy drugs."

"But they have side effects!"

The skeptical newcomer is correct. Drugs do have side effects.

For starters, side effects can be either beneficial or harmful. An example of a beneficial effect is aspirin; originally developed as an anti-inflammatory drug and a painkiller, aspirin has serendipitously found an alternative use as a "blood thinner," a drug that prevents clot formation.

DOI: 10.1201/9781003346661-16

The term, side effect is often used synonymously with *adverse drug reactions (ADRs)*, which are unpleasant or potentially harmful effects caused by drugs. They can vary from minor problems like a runny nose to life-threatening events such as a heart attack or liver damage. To my knowledge, every drug has at least one or two adverse reactions listed on the package insert; the frequency of these effects can range from very rare to frequent. Let's consider a few examples.

A drug that is designed to, say, lower one's blood pressure may cause headache and dizziness in some patients because of a sudden drop in blood pressure. This is easily comprehensible—the symptoms reflect an exaggerated effect of the intended use, an augmentation of the drug's pharmacological action. Such reactions are highly dependent on the dose, they are also predictable. Most ADRs belong to this type.

Other drugs may also cause adverse effects, but these effects have nothing to do with the drug's pharmacological, intended action. Instead, it is a toxic response that can be explained by the chemical properties of that particular drug or its metabolites. Importantly, like the example above, such adverse reactions are dose-dependent and predictable. An example is liver failure induced by excessively high doses of the painkiller, acetaminophen (we have covered this in Chapter 6). For lack of a better term, such ADRs have been termed Type-A reactions (see Table 14.1).

In contrast, another kind of adverse drug reaction can occur out of the blue, in an unpredictable way, and in very few patients only, while the overwhelming majority of recipients tolerate the drug well.

Table 14.1   The Two Types of Adverse Drug Reactions

ADVERSE DRUG REACTIONS	
TYPE A (INTRINSIC)	TYPE B (IDIOSYNCRATIC)
Rare to frequent	Very rare
Predictable	Unpredictable
Dose-related	Dose relationship not obvious
Identified early in animal and/or human studies	Often recognized post-marketing only
Toxic response can be explained by toxicodynamics or -kinetics	Unclear mechanism; often immune-mediated
Associated with specific *chemical*	Associated with individual *patient* (+ context)

Note: The predictability of unpredictable reactions has become considerably better in recent years. Genetic factors are increasingly known to be a determinant of individual susceptibility.

The reaction cannot be readily explained by the drug's intended purpose (it's not an exaggerated pharmacological effect), nor by the drug's chemical characteristics. In addition, the dose relationship is not easily recognizable (although there must be some sort of dose dependence, like for every biological process). It is not so much the chemical itself as a combination of the individual patient's genetic and acquired features and the context (other medications, underlying disease, etc.) that can trigger such a strange and unexpected ADR. This type of reaction is called Type-B reaction, or *idiosyncratic drug reaction*. An example is a severe skin rash triggered by, say, a certain antibiotic.

Here's a frequently asked question: Why isn't it possible to detect and prevent such adverse drug reactions *before* patients are being harmed? Aren't there any testing protocols in place before a drug is put on the market?

And here's the short answer (flashing ahead to the next chapter): most adverse effects are revealed long before a chemical becomes a real drug. In fact, the toxicity of lead compounds and drug candidates is frequently recognized early in cell, organ, or lab animal studies, or later in the clinical phases of drug development (in patients). In other words, such chemicals are being eliminated—the earlier the better. In fact, the vast majority of all chemicals that are investigated for their potential use as a new drug are dumped and abandoned without the average consumer hearing about it (stay tuned for more). For those few approved drugs that make it all the way to the market, ADRs can be prevented provided the patient complies and takes the proper dose. But here's the exception: the rare but feared idiosyncratic drug reactions are often detected too late, surfacing only when millions of patients have already been treated with the new drug. When this happens, and if the health risk is unacceptable because it outweighs the drug's benefit, the drug company will pull that drug from the market—something that has happened in the past (see Chapter 15). However, with the development of more and more sophisticated testing programs, such dead-end drugs fortunately have become very rare, and drugs have become much safer.

(For pharmacy practice fanciers: so far, we've been talking about single drugs, but what about the risk of adverse reactions when several drugs are prescribed and taken at the same time? This so-called *polypharmacy* happens quite frequently, especially in aging populations. For example, it is quite common for patients with cardiovascular problems to take, say, one to three different drugs to lower the blood pressure, another one to reduce the "bad" cholesterol levels, another one to prevent blood clotting, maybe some vitamin supplements, and whatnot. There are some dazzling combinations of multiple (>5) drugs. Why do we need to pay attention here? Not only may the risk for multiple adverse reactions increase, but different drugs may also interact with each other in strange ways, and accurate predictions become tricky. The larger the number of different drugs taken together, the higher the chance that the risk may be greater than the benefit. Cutting back ("deprescribing") is the solution.)

Back to our cocktail party.

"Drugs? Safe? Ha, my ass," a smug guy who must have overheard our arguments throws in. More people are joining the discussion group. "How about painkillers? Last year, more people died from opioids than from Covid-19." Sadly, that's true.

"It's all about exposure," you repeat, a harried look on your face. "You cannot prevent someone from taking an overdose of something—anything—and then make the company responsible for it. Every drug is potentially toxic."

That's a rather controversial topic. You can tell by the look on their face; some folks turn away, seeking a more enjoyable conversation than chatting about the off-label use of a drug.

"Depends on the margin of safety," a man with dark hornrims says. "Especially for opioids." You bet he's a pharmacologist.

Suddenly you realize you're alone with him. Interesting conversation, but not a cheerful party topic.

This transitions right to the next section.

The Opioid Crisis

The opioid crisis (also called the opioid epidemic) is a medical, social, and personal crisis describing the overuse and misuse of opioid drugs, a class of medications prescribed against severe acute pain. The word "crisis" is appropriate; sadly, in the past two decades, several hundred thousand Americans have died from improper use of opioids.

The problem is based both on the opioid drugs' high potency and their addictive potential. Although it might be tempting to disparage the chemicals, it is not the opioid drugs themselves that are to be blamed. Opioids can effectively relieve severe pain and are therapeutically beneficial—when used judiciously. They have been approved for medical use and thus cannot simply be banned. It is their misuse, abuse, and, of course, illicit synthesis and distribution that is the core of the problem.

Opioids are a large family of chemical compounds that occur either naturally (e.g., morphine) or are synthesized (e.g., oxycodone, hydrocodone, or fentanyl). (The name "opiates" is an older term for drugs derived from opium.) They are used as strong painkillers (analgesics) and anesthetics. To elicit their effects, opioids bind to one or several receptors in the nervous system, mostly the so-called μ (mu)-opioid receptor. Their potency on that receptor greatly varies; compared to morphine, fentanyl is approximately 100 times more potent, and sufentanil even 1,000 times. Carfentanyl is so potent (and therefore dangerous) that it is only used in veterinary medicine (and for tranquilizing elephants).

As we all know, opioids have quite a few typical adverse drug reactions caused by the activation of the μ-opioid receptor (involving a complex interaction with other receptors as well): sedation, pinpoint pupils, nausea, vomiting, constipation. Much feared, however, is severe respiratory depression (the patient stops breathing)—the major cause of death after an overdose. Fortunately, in the event of an acute overdose of an opioid, there is an antidote that can save lives if given quickly enough: naloxone (Narcan). This antidote avidly binds to the same μ-opioid receptor as the dangerous opioids, thus displacing them from the receptor and blocking their dangerous effects.

The known dangers inherent in opioids are two-fold: first, opioid use can induce *tolerance*, requiring ever-increasing doses to elicit the same effect, and, second, it can cause *dependence*, producing severe withdrawal symptoms after discontinuation, leading to addiction. That former feature can be especially dangerous in the following current example.

Opioids are often combined with other, non-opioid painkillers (e.g., acetaminophen) to minimize the dose, and therefore reduce the well-known undesirable adverse effects. As mentioned earlier, acetaminophen is safe when used at a therapeutic dosage. However, because the pain-relieving efficacy of the opioid/acetaminophen combination gradually dwindles because of the above-mentioned tolerance effect for the opioid, the patients often proceed to take more pills to achieve the same effect as before. Sadly, this has resulted in avoidable cases of liver failure, some fatal—not because of the opioid, but because of an excessively high dose of acetaminophen.

There is no easy recipe of how to bend the curve, reverse the upwards trend, and get out of the opioid crisis. However, great efforts are being undertaken to improve prescription practices for opioid analgesics, educate the public, identify populations at risk, and improve intervention strategies. The search for new medicines and new targets is ongoing, but it looks like opioid analgesics will be around for quite a while. They have proven to be extremely useful and helpful for patients. Again, the problem is not the chemicals—managing their proper use and dosage has remained the main challenge.

People Are Different—Individual Susceptibility

Not everybody reacts to a pharmaceutical drug the same way. People are different. Not only with respect to age, gender, race, underlying chronic disease, etc. but also to certain genetic (and acquired) traits that render certain individuals particularly sensitive—or resistant—to drugs. This is the basis for "personalized medicine," i.e., tailoring the treatment of a patient to the person's individual characteristics.

Let's look at an example.

Harking back to Chapter 6, we know that the liver and other organs have the capacity to metabolize drugs and other chemicals that enter the bloodstream. In most cases, these altered chemicals become better soluble in the blood plasma and can be more easily excreted. Responsible for this chemical modification is a bunch of specialized proteins (aka drug-metabolizing enzymes) that can grab one individual molecule of the foreign chemical, modify it, then release it, then get prepared to grab the next molecule, and so forth. Over time, there will be fewer intact parent molecules of our drug but an increased amount of metabolites present. Consequently, the administered pharmaceutical drug gradually loses its activity because its concentration in the body keeps diminishing. The therapeutic effect is gradually lost—until it's time for the next dose (e.g., 24 hours later).

The dosage of a pharmaceutical is carefully chosen to make sure the drug stays in the body long enough to exert its beneficial effect. However, if the drug-metabolizing enzyme works too efficiently in a patient, the drug will be metabolized and cleared too fast, and there is no real therapeutic effect. On the other hand, if the enzyme works too slowly, the drug cannot be efficiently eliminated and stays in the system for too long, eventually even reaching levels in the toxic range.

One of these drug-metabolizing enzymes in the liver is a protein called CYP2D6 (for toxicology mavens: one of the many human cytochrome P450s). It is responsible for metabolizing approximately 25% of all known drugs (including opioids, beta-blockers, antidepressants, and many others). It is known that the gene in our DNA that encodes for CYP2D6 is present in people in different variants; in some folks that gene is entirely missing, in others, it produces a highly active enzyme, still in others the gene variant encodes for a CYP2D6 protein that features intermediate activity. As a result, different people metabolize those drugs for which CYP2D6 is competent at different rates; there are the "ultra-rapid metabolizers," the "intermediate metabolizers," and the "poor metabolizers." What are the implications for the average patient who must take such a prescription drug? It means that the intermediate metabolizers (curve A in Figure 14.1) will have

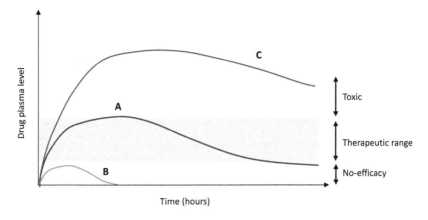

Figure 14.1 Time course of drug plasma concentrations for individuals with different rates of drug-metabolizing enzymes. A, intermediate metabolizer; B, ultrarapid metabolizer; C, poor metabolizer.

drug plasma levels well in the desired therapeutic range. In contrast, ultrarapid metabolizers (curve B) will have low levels and a very small AUC (area under the curve, see Chapter 3). They probably will not feel any effect of that drug. Finally, the poor metabolizers (curve C), who do not clear the drug efficiently, will inadvertently get a massive overdose that can go far into the toxic concentration range.

Translated into the situation of an opioid, this would mean that an ultra-rapid metabolizer would get little pain relief, whereas a poor metabolizer would suffer from dangerous toxic effects after the same average, "therapeutic" dose.

Such so-called *genetic polymorphisms* (different variants of the same gene distributed in a population) exist for other critical proteins as well. They are one of the factors responsible for how different people handle chemicals differently and how differences in toxicokinetics can influence the risk after exposure to the same dose of a chemical.

Differences in the toxic response from one person to another may have yet other underlying reasons. Their drug metabolism capacity may be the same, but they may differ in the extent of their defense capacity against toxic chemicals or their metabolites. For example, they might be less capable of fighting off increased oxidant stress (you may want to revert to Chapter 8) because some of the antioxidant defense systems might be compromised. That is exactly the case with one of the most frequent genetic polymorphisms in humans, affecting hundreds of million people worldwide: the G6PD polymorphism. Here's what it is, and what it means.

Some people feature one of several genetic variants in a gene called glucose-6-phosphate dehydrogenase (G6PD), resulting in a functional deficiency of the protein that is normally encoded by that gene. The G6PD protein is not completely absent but has some residual function (if it were fully deleted it would be deadly for that individual). G6PD is critically involved in the production of glutathione in red blood cells, and red cells are normally packed with glutathione, to protect them from oxidant stress. Red cells in people with G6PD deficiency have insufficient levels of glutathione and are therefore vulnerable to attacks from reactive metabolites and oxygen radicals. One drug that produces both reactive metabolites and oxidant stress is the antimalarial drug primaquine. Primaquine is well tolerated in people with normal levels of glutathione in their red cells but causes toxicity to red cells in people with G6PD deficiency—their hemoglobin in the red cells gets damaged, so that the now poorly functioning red cells are removed from the circulation prematurely and recycled. G6PD deficiency is not rare; in fact, the genetic variations (there are many forms of it) are particularly frequent among African-Americans (about 11%), and in parts of the Mediterranean the frequency is almost at 50%.

The two examples discussed above are relevant because they cause problems for patients, but they have one great advantage: the adverse drug reactions can be predicted (by analysis of the respective genes in the DNA). In other words, one can find out who, e.g., is a poor metabolizer of CYP2D6-dependent drugs, or who is a carrier of the deficient G6PD gene, and therefore will have a greater risk for developing drug toxicity.

Unfortunately, there are other cases where an adverse reaction may occur out of nowhere; unexpected, and unpredictable, as mentioned in a previous section of this chapter. The reasons why such rare, unpredictable toxicity of a normally well-tolerated drug occurs in some individuals only, but not in others, are mostly unknown. However, evidence shows that in many cases the immune system is involved. Fortunately, these reactions are rare (in the range of 1:10,000 to 1:100,000 individuals). Echoing the section on ADRs (above), such rare and unpredictable drug reactions are termed *idiosyncratic*. Triggered by a drug, a person's immune system can go haywire and attack an organ (e.g., the skin). We are only beginning to understand some other (non-immune) mechanisms that are the underlying causes of other rare and bizarre forms of drug toxicity. The most affected organ is the liver, followed by the cardiac system and the kidneys.

Again, it is obvious that due to the low incidence of such idiosyncratic drug reactions the problem becomes only apparent after several thousand or even millions of patients have been exposed to a drug. This is usually the case *after* the launching of a new drug on the market; in fact, sometimes it takes several years until such a safety issue becomes evident. When this occurs, the pharmaceutical company that sells the drug has a real dilemma. It has invested millions of dollars in the development, made sure the drug works and is safe at the right dosage, finally gets FDA approval, and now has a great new drug that helps treat diseases in patients—only to be confronted later with the tough decision of whether to keep it on the market or else withdraw it due to the emerging idiosyncratic toxicity in a few patients. Initially, such a drug will receive a *black-box warning* (a warning of a serious risk on the package insert, framed in a black box, therefore the name). If the risk outweighs the overall benefit, e.g., when deaths occur because of an adverse reaction, the company will decide to withdraw the drug from the market. This has happened occasionally in the past. To do so is a tough decision and, among many other factors, the severity of the disease itself (the indication) will be considered: a headache pill obviously is different from a badly needed antibiotic or cancer therapeutic. Needless to say, this is feared by any company, and great efforts have been underway to prevent such rare incidents from happening again—with success.

So, taking the individual susceptibility factor into account, the risk equation that we began to construct earlier can be further expanded see Figure 14.2.

Taken together, different people react differently to drugs because our genetic and acquired characteristics are all different. Pegging drugs as "highly toxic" just because they may cause adverse drug reactions is unjustified. Untreated diseases can be deadly too. There can always be rare, unpredictable toxicity of drugs—and let's not forget that the vast majority of potential health risks are being picked up during the development of new drugs. How this is done is the topic of the next chapter.

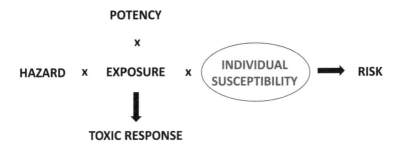

Figure 14.2 Role of individual susceptibility factors in determining a chemical risk assessment.

TAKE-HOME MESSAGE

- All pharmaceutical drugs can cause adverse drug reactions, but they are usually rare, can be explained, and depend on the exposure.
- ADRs must be balanced against the benefit of that drug.
- Very rare, unpredictable adverse drug reactions can also occur, not because of a "bad" drug, but because of a unique sensitivity of individual patients.
- Opioids can be dangerous when inappropriately used because of their high potency and because they can induce tolerance and dependence.
- Different people metabolize drugs at different rates, with implications for health risks. Other individuals differ with respect to how efficiently they can fight off a harmful threat from a chemical. Personalized medicine is grounded on factoring in these individual differences.
- *Specially geared*: If you know of a drug that has no adverse effects whatsoever, chances are that it's not very effective, except for its placebo effects.

PART III
THE RISK

15
SAFETY ASSESSMENT

Can We Predict and Prevent a Toxic Response?

One of the hardest things is to predict exactly what types of adverse reactions a new chemical might trigger in humans, and how frequently that will happen in populations. Unfortunately, the pharmaceutical and agrochemical companies don't have a crystal ball (although they could save a lot of money and avoid the hassle if they had one). We've had big surprises in the past, but similar rude awakenings are getting rarer as our predictive tools are getting better and better.

"Why can't they finally come up with a drug, or a pesticide, or an industrial chemical, that is devoid of adverse effects?" We have all heard these and similar frustrated comments. Meanwhile, you know that this is not possible because we are complex human beings, and everything will induce a toxic response at some point as the exposure keeps ramping up. Our regulatory authorities must simply define the borderline between a reasonably low risk, in everyday situations, and an unacceptable increased risk at higher exposure.

"Yeah, I get that," a resentful consumer might complain. "But even the experts don't know what they're talking about. Take that pesticide, for example. After all these years, they're still arguing back and forth whether it may cause cancer. When are they finally proving that it's not gonna happen, that it's safe?"

This is more of a philosophical than a scientific question. Harking back to Chapter 2, safety is a relative term. Can one ever predict that something is *not* going to happen? Obviously not. At the most, one could say that the risk of something to happen is low. Contrary to popular belief, a company is not obliged to demonstrate that a new chemical is safe (in fact, they wouldn't be able to do so). It is not about the absence of toxicity. Instead, they must provide evidence for a dose-versus-toxic-response relationship.

To that effect, they must follow rigorous testing protocols. A company must strictly adhere to an entire battery of internationally standardized tests when it wants to launch a new product. They must then submit an application to the regulatory authorities, and the product

DOI: 10.1201/9781003346661-18

can only be put on the market when it is clear what types of toxic responses, and to what degree, are to be expected under the conditions outlined.

Let's focus first on pharmaceutical drugs.

Drug Development

To develop a new drug and get it approved is not only expensive, but it also takes time—twelve to fifteen years on average (although the time can be shortened in special cases, e.g., for emergency use authorization). The reason is, it is hard enough to discover a new chemical that has a certain desired effect, but it also must be demonstrated that the drug has a low risk and is safe for its intended dosage and use in patients—or, in other words, that the benefit outweighs the risk. In addition, an in-depth understanding of the disease for which a drug is used is also required, and there are still some diseases that are enigmatic.

All this must be accomplished by testing the chemical on different levels. The first phase is called the *drug discovery* phase (see Figure 15.1). First, a bunch of chemicals are screened for the desired efficacy, e.g., how strongly they bind to a certain receptor, or how specifically they inhibit a certain biochemical function that plays a role in the targeted disease. This is not done in live organisms, but rather

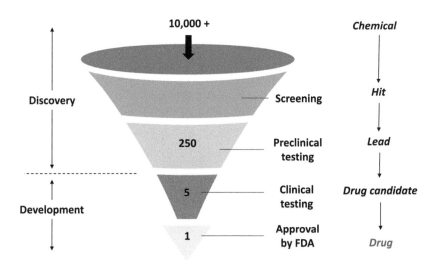

Figure 15.1   Attrition of chemicals and drug candidates in drug development.

performed *in vitro* or *in silico*, allowing for a stunningly high throughput rate, i.e., testing many different chemical compounds in a short time—for more, keep going.

Next, the new compound's efficacy must be demonstrated in animals that ideally model the disease for which the drug is intended (e.g., diabetic mice for testing an anti-diabetic drug). At the same time, the toxicity profile of the new chemical must be assessed by validated tests. These tests must follow internationally standardized protocols, performed under conditions of so-called *Good Laboratory Practice (GLP)*, enabling data reproducibility, reliability, data recording, and reconstruction of the studies. In addition, it guarantees the integrity of the studies and optimal maintenance of animal facilities and the welfare of laboratory animals, with rigorous checks and balances. Again, the aim of these studies is not to give a sigh of relief and say, "everything's fine, there's no tox"—not a good idea to submit that dossier as a new drug application to the FDA. Instead, the purpose of those *in vivo* studies (i.e., in live organisms) is to demonstrate at what dose levels a toxic response will occur, in what target organ, and what type of response that is. Up to this point, no humans were involved for obvious ethical reasons; therefore, these studies are called *preclinical* studies (because they precede the first clinical trials).

The final but also most challenging phase is called the *drug development* phase as from now on humans will be exposed (see Figure 15.1). During this clinical testing, both the efficacy and safety of the new drug must be assessed in patients, again following rigorously standardized and independently vetted protocols that must adhere to the so-called *Good Clinical Practice (GCP)*. This is an international quality standard that delineates the ethical guidelines and provides the technical and scientific framework, making the design, recording, and reporting of human trials transparent for the public.

There is a long way to go from screening an enormous number of different chemicals for a potential "hit," which may give rise to a promising "lead" compound, resulting in a drug "candidate" that may or may not wind up to become a marketed drug. Not only are the expenditures of time and money huge, but also the process is stinted by attrition. Typically, for some 10,000 or more chemicals that are being screened initially, only some 250 will undergo preclinical testing, leading to approximately 5 clinical candidates, and resulting in one (1) marketed drug. Why is the attrition rate so high? Most chemicals fall by the wayside because of a lack of efficacy and/or safety issues. That single drug that comes out at the lower end of the development "funnel" (see Figure 15.1) has been rigorously tested for years and put through the acid test in numerous ways. How that happens will be explained in the following section.

One thing is clear, though. If a lead, or a drug candidate, doesn't make it at one point or another along the pipeline, it is best if the failure is detected as early as possible. This would not only save valuable time and prevent wasting money, but it could also avoid unexpected toxic responses in humans. Realizing that a drug poses an unacceptable risk late during the clinical trials, or, even worse, in the post-marketing phase, is a disaster—not only for those patients affected but also for the company. A promising therapeutic for human disease would recede into the distance.

Translation from Mice to Humans?

We often hear claims that results obtained from treating mice or rats with a drug are irrelevant for humans and that, therefore, all *in vivo* studies with laboratory animals should be abandoned. When taking a closer look, though, these are two different things that need to be addressed separately. The first half of the sentence is simply not true and reflects ignorance of or unwillingness to accept some of the most basic biological facts. The second half of the sentence is not true as it is based on a wrong premise. The timely and topical issue will be discussed in detail below—just one thing given away here: over the past decades, both the regulatory agencies and drug companies, as well as the research scientists themselves, have made enormous efforts to reduce the number of animals wherever possible, and the trend continues. Nota bene, without compromising on safety.

So why mice and rats? Because human data are scarce or, in the case of new chemicals, entirely absent, and because one cannot use people as guinea pigs and expose them to unknown chemicals for ethical reasons, one must fall back on "animal models" as surrogates. The choice of the right animal model is crucial: the wrong model would be irrelevant or even deceptive and dangerous for humans—therefore, the relevance of a specific animal model must be established early.

Laboratory mice and humans are both mammals and, despite the many obvious differences, have a lot in common. They share many anatomical, physiological, and biochemical features. There are numerous examples where a rodent model is ideal for studying the efficacy and safety of a drug and where such a model is predictive of the human situation. For example, the painkiller acetaminophen (discussed in Chapters 6 and 8) has been extensively studied in the mouse with regard to its metabolism in the liver, how the toxic metabolite is trapped by defense mechanisms including glutathione, and how exactly, on a molecular level, an overdose of acetaminophen can harm the liver. It has been generally accepted that the mouse model is excellent to study the toxicity of acetaminophen, reflecting exactly what's going on in patients.

On the other hand, nobody would deny that there are examples where a mouse or a rat is a poor choice to study a specific toxic effect. For instance, the anti-cancer drug tamoxifen, widely used to treat breast cancer, produces liver cancer in rats by a genotoxic (DNA-damaging) mechanism. A careful analysis of how tamoxifen is being metabolized in rats, however, revealed that its metabolism in rats is different from that in humans. (For toxicokinetics enthusiasts: rats produce a reactive metabolite that is likely responsible for the DNA damage, whereas humans generate a harmless metabolite that is readily excreted. Indeed, there is no indication that tamoxifen causes liver cancer in women.)

You are probably wondering how on earth one can find out whether in a new situation a mouse or a rat would be the better model for the human situation. One way is to generate preliminary data with cell culture models or organ cultures from a variety of species and then go from there.

Non-Clinical Studies

Suppose we have a promising new lead compound. We have learned how it is being metabolized in liver cells and what kind of effects it might have in other cells—but will it work *in vivo*? It's time to move on to non-clinical studies in laboratory animals. (So far, we have been using the term "preclinical" to describe the phase preceding the clinical phase. However, as of now, "non-clinical" is the preferred term because extended animal studies will continue and overlap with the clinical trials.)

Unlike the frequently misrepresented facts, non-clinical studies in animals are not simple and brutal experiments where a bunch of mice are injected with the new chemical and the dead bodies counted the next morning. Those dark ages are long gone. People who still believe that's the way an overall safety assessment is performed have a fundamental lack of knowledge of the current refined possibilities of how to test chemicals, and they are unaware of the strict requirements and meticulous guidelines set by the regulatory authorities.

Animal studies can only be initiated after a critical review of the protocols by independent expert committees. The aim of the studies, the number of animals, the exact procedures, and efforts how to minimize distress and optimize the care of the lab animals must be documented and justified in detail.

Before we delve into the details, let's get something straight—the question of how to determine acute toxicity. *Acute toxicity*, by definition, comprises the adverse effects of a chemical after one single dose (or daily doses for no longer than one week). The classical test, still used in some countries and frequently cited in textbooks, is the so-called LD_{50} *test* (you may remember the term from Chapter 2). This value is the median lethal dose of a compound, i.e., that dose where 50% of the animals die. This test is not only outdated and ethically questionable but also inaccurate. The results vary greatly from one study to another and provide no other information than lethality. For these reasons, the regulatory agencies in the United States (FDA, EPA) and Europe (EMA, EEA) no longer require LD_{50} tests, and other countries have banned it too. There are better methods available to estimate the acute toxicity of a compound, using a much smaller number of animals, and not always using lethality as an endpoint.

After having sorted this out, let's move on to the assessment of chronic toxicity. *Chronic toxicity*, again by definition, reflects the sum of adverse effects after repeated administration of a chemical for a prolonged period. Typically, for rats, this is a 3-month study (equaling about 10% of the animal's lifespan). The aim of these studies is not to show that there is no toxicity and that, therefore, the compound is probably safe (we should meanwhile understand that this would be futile and inconclusive) but rather to define a dose level at which no toxic effects are observed (*no-observed-adverse-effect-level, NOAEL*). Furthermore, a potential target organ (liver, kidney, lung, blood cells, etc.) must be determined at higher doses, and, importantly, a dose level found for the subsequent long-term studies and a basis for the calculation of the first-in-human dose (stay tuned). These are sophisticated studies using two different species, both sexes, with at least three different dose groups (low, medium, and high dose), and a control group receiving exactly the same treatment except the drug. The highest dose should elicit some sort of toxicity but must not be lethal. A plethora of "clinical-chemical" data are collected over time, including markers of organ function, urine analysis, and blood work—many of the methods similar or identical to those being used in a clinic for monitoring patient health. In addition, and most importantly, at the end of the study, microscopic tissue sections of every organ of every animal in the study are evaluated by specialized and highly trained veterinary pathologists.

Again, and this can't be stressed enough, it is the governmental regulatory agencies that require these studies prior to the approval of a new drug. Should there be an unexpected significant finding in one organ or another, the companies must conduct additional studies aimed at finding a valid mechanistic explanation for the changes.

Other elaborate studies include investigations on potential effects of a drug candidate on the offspring, on the development before and after birth, and, importantly, the ability of a chemical to induce gene mutations (DNA changes). Specially trained scientists work in teams to put together a comprehensive data package of all these effects.

Finally, to complete the non-clinical safety assessment, studies to investigate the potential of a chemical to induce tumors (benign tumors or cancer) must be performed (you might want to turn back the pages to Chapter 7). Two biological facts make these "carcinogenicity studies" especially complex. First, the risk for tumor formation is low in young individuals but increases with age (both in animals and humans). Second, there may be a long latency time from exposure to a carcinogenic chemical to the manifestation of a tumor (in humans some 20–25 years). Therefore, cancer studies must be designed to last for almost the entire life span of lab animals—the average life span of a rat is 2–3 years. Again, every organ of every animal is scrupulously scanned for any tumors and checked against those in the control group to correct for "spontaneous" tumors.

In recent years, though, the criticism among scientists about the validity of such lifetime bioassays for carcinogens has become louder. It is obvious that correlations between the cancer-inducing potential of a chemical in animals and the human risk, when made on a scientific basis, are extremely difficult. Strict regulatory guidelines, imposed as a matter of law, are based on certain default assumptions that have become somewhat outdated. For example, the premise that extrapolation from lifelong, high-dose exposure in animals to short-term, low-dose exposure in humans is linear, is scientifically not tenable (more on this in Chapter 17).

Clinical Trials

Following the initial non-clinical studies, a drug candidate enters the clinical trial phase. This is typically done in a tiered approach (called Phase I through IV). Human studies can only be started after a rigorous review of all protocols by experts and must be approved by ethical committees. Clinical trials typically continue for several years and are the most expensive part of the long journey to approval and marketing of a drug.

Phase-I trials are performed to assess the safety of a new drug and to set the dose (the point is to pick up any untoward effects early). The studies are done on healthy volunteers, normally about 50 individuals, and proceed under strict medical surveillance.

In Phase II, the drug is given to a small number (up to 300) of carefully selected patients having the disease for which the drug is intended. The major goal is to assess the efficacy of the drug (whether it works in humans).

Phase III comprises a large number of patients (1,000 to 3,000). The major goal is to further confirm the drug's efficacy, identify potential adverse effects, and compare it to commonly used other drugs. These drug trials are designed as *randomized, double-blind, placebo-controlled trials*. What does that crabbed term mean in plain language?

Placebo-controlled means that a drug group is run against a placebo group (a harmless "sugar pill" group) to exclude a non-specific effect. Randomized means that each participant is randomly assigned to receive the new drug or a placebo (control). Double-blind means that neither the participant nor the doctor knows who is receiving the real drug or the placebo—to exclude any bias.

Finally, Phase IV follows the approval and launching of a new drug (it's also called the post-marketing surveillance phase). The purpose is to confirm efficacy and safety in a very large number of patients and to record any untoward effects that are so rare that they would only become apparent after thousands (millions) of people are being treated (see Chapter 14).

First Exposure

Here's another typical and embarrassing situation, happening around a fictional dinner table on an ordinary Saturday night.

"That's unethical, totally unacceptable," one of the guests pipes up. "Using human volunteers as guinea pigs!" I wish we hadn't started this explosive conversation topic. I don't even know why it started.

"What do you mean?" I already smell the rat.

"Those new drugs that Big Pharma is developing. First, they poison mice, then they force a patient to swallow a megadose of that stuff and see what happens."

If that were the current approach for clinical trials, no drug company could last very long. I'm sure our guest knows that as well.

"Someone has to be the first person to take a new drug, there's no way around it," I say, realizing too late that this may sound sardonic, which is not my intention. "I mean, they have a pretty good idea where to start and what to expect."

"Would *you* want to volunteer for that?" The guy jabs a finger at my chest. "It's Russian roulette. Could be a deadly dose."

"You have any idea how the first-in-human dose is calculated?" Someone else comes to my aid.

Our truculent guest looks non-plussed.

"What she means is, it's a matter of how much," I explain. "It's all about exposure."

In fact, it's quite a tricky question. For a new chemical, how does one arrive at a safe starting dose for humans? Answer for pundits: by allometric dose scaling. For newbies: here's what's behind it.

We have learned that we can obtain a NOAEL (a no-observed-adverse-effect level) from a mouse or rat study, but how do we translate that into the first human dose?

You have certainly noticed that, up to this point in the book, a dose of a chemical has always been expressed as a certain amount per body weight (e.g., mg/kg body weight). However, when you want to translate (extrapolate) a NOAEL dose obtained from laboratory animals into an equivalent human dose, taking the body weight as a reference is, strictly speaking, not accurate. Why not? Because of the obvious difference in body mass between a tiny mouse and a burly human. It may sound confusing, but it is the *body surface area*, rather than the weight (mass) of an animal, that reflects its overall physiologic metabolic rate. In other words, the larger the animal, the slower the general physiological processes. A mouse (weighing approximately 20 g) burns more calories than a rat (approximately 200 g), and even more than a human (average 70 kg) if expressed on a body weight basis, but about the same, if expressed on a body surface area basis. The same holds true for the metabolism of a drug; therefore mg/kg body weight must be transformed into mg/m^2 body surface. To do so, i.e., to calculate a human equivalent dose, a mouse dose (NOAEL) should be divided by roughly a factor of 12, and a rat dose divided by a factor of 6. This is summarized in Figure 15.2. Finally, a safety factor (better: uncertainty factor) must be applied (usually a factor of 10), which will give an estimate of the first human dose.

Our flustered guest nods in agreement and decides to attack the pork chop on his plate. Thankfully, and elegantly as ever, my wife shifts the conversation to a lighter note.

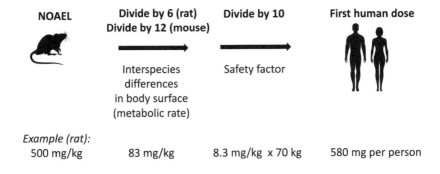

NOAEL	**Divide by 6 (rat)** **Divide by 12 (mouse)**	**Divide by 10**	**First human dose**
	Interspecies differences in body surface (metabolic rate)	Safety factor	
Example (rat): 500 mg/kg	83 mg/kg	8.3 mg/kg x 70 kg	580 mg per person

Figure 15.2   Calculation of the first human dose for a new drug.

Note: The body surface area for an adult human is approximately 1.6 m^2, for a rat 0.025 m^2, and for a mouse 0.007 m^2). The safety factor of 10 is used as an arbitrarily chosen buffer.

Testing of Agrochemicals

The major difference between the use of pharmaceutical drugs and agrochemicals (pesticides) is the obvious fact that the latter are not intended for human exposure. Instead, the primary goal of testing agrochemicals is, besides protecting crops and human health, to avoid toxic effects on the environment and ecosystems. Exposure to humans will primarily be limited to ingesting or inhaling or touching residual amounts of pesticides in foods, in the air, or in the soil. Occupational exposure during manufacturing pesticides, and spreading them on the field crops, of course, poses a much greater risk.

Overall, non-clinical testing of new agrochemicals follows similar guidelines and regulations as the ones in place for pharmaceutical drugs. A major effort is made to detect the potential endocrine-disrupting activity of agrochemicals, as well as neurotoxic effects, two well-recognized potential targets (a topic broached in Chapter 7). Human clinical trials are not part of the overall assessment package and are not required by the regulatory agencies, which makes sense.

Computer Simulations, Omics, and Organs-on-a-Chip

Despite tremendous efforts to continuously improve the existing non-clinical testing methods and further advance clinical development programs, unexpected adverse effects of chemicals have persisted, entailing regulatory measures. However, novel approaches and refined techniques have been developed, resulting in better prediction and a greater reduction of such adverse reactions occurring out of the blue.

A detailed discussion of these emerging new models is beyond the scope of this book, but some of them will be briefly mentioned. One important modern area is *systems toxicology*, a widely used term that is not easy to grasp. Systems toxicology is the quantitative integration of a myriad of data from *in vivo* and *in vitro* studies, as well as *in silico* experiments (computer-simulated predictions of effects based on the structure of a chemical). An important tool is the application of so-called "-omics" techniques, e.g., *proteomics* (a simultaneous quantitative and qualitative determination of hundreds or thousands of proteins in a sample), or *transcriptomics* (measuring the different RNAs in a sample, i.e., the direct transcripts from different activated genes in the DNA).

Importantly, all these data must be linked with drug exposure—something that is hard, if not impossible, to do for a single person; only computer models can come up with creating and analyzing the resulting complex networks.

Another rapidly growing area has been cell or organ cultures. One disadvantage of traditional cell culture models is that they are static systems, i.e., they do not consider blood circulation that brings in new supplies with nutrients and oxygen and removes waste products. One way to circumvent this is the development of so-called "organ-on-a-chip" (or organoid) models, i.e., three-dimensional cell cultures, often combining several different cell types, embedded in a microenvironment in a microfluidic device, mimicking in many ways a real organ. This sophisticated technology, which has already been established for many organs, will undoubtedly become an important tool in drug safety evaluation.

There is one caveat, though. Cell and organoid cultures, although widely used, have clear limitations when one wants to model complex processes that involve the concerted processes of the whole organism. How is one going to study, say, the effects of a drug on blood pressure in a cell culture dish? A similar difficulty is to evaluate a drug's effects on the nervous system. How can one measure functional or behavioral changes? Animal models will likely continue to be used when there is no other option.

Despite the cries of naysayers, lab animals are still being used. However, within the scientific community, great efforts have been underway to minimize their use and at the same time maximize the information from such studies. One way is to apply the so-called "3R" principle, which stands for refining, reducing, and replacing the use of animals (responsibly, would be the fourth R).

The goal of the science of toxicology is to model, understand, predict, and ultimately prevent toxic responses from exposure to chemicals. There is still a long way to go, but we've made tremendous progress.

After the enormous task of safety assessment for a chemical has been completed (in essence, a reliable dose-response relationship established), the next question pops up, and it goes now to the regulatory agencies: where are the red lines, the "safe-exposure" limits for all those potentially dangerous chemicals, and how do we know where to set them? This will be explored in the next chapter.

TAKE-HOME MESSAGE

- It takes 12–15 years to develop a new pharmaceutical drug from scratch. On average, only 1 out of 10,000 or more chemicals that are screened will eventually make it to the market.
- Preclinical (non-clinical) studies in laboratory animals comprise extensive standardized testing protocols for both acute and chronic toxicity and toxicokinetic studies. They include organ function and toxicity tests, blood work, a full pathology investigation, and carcinogenicity studies.
- Clinical trials start with a carefully estimated (low) dose in healthy volunteers, followed by administration of the new drug to a few, then to a larger number of patients to assess efficacy and safety. However, rare adverse drug reactions may still occur after approval, in the post-marketing phase with millions of patients receiving the drug.
- *In vitro*-studies with cell cultures, organs-on-a-chip, omics techniques to detect specific genes being turned on or switched off, and computer simulations will be increasingly applied. However, studies with a suitable animal model are still indispensable currently although the future vision is to get rid of traditional large-scale animal studies.
- The tragic "accidents" that happened in the past are not proof of the inadequacy of non-clinical studies. Rather, their occurrence was possible because we did not yet have the sophisticated and required battery of testing protocols that are currently in place.
- The "3R" principle (refine, reduce, replace animal studies— with responsibility) has been increasingly applied in many countries.
- *For good measure*: At a dinner party, think twice before embarking on a heated, emotional debate on politics, religion, or drug development.

16

ACCEPTABLE LIMITS, TOLERANCE, AND RED LINES

Crossing the Line

Another one of those ominous headlines

You're reading a report that in your neighboring town, the threshold level for a certain hazardous chemical in the drinking water has been exceeded last week—again. By almost 10%. All alarm bells go on—and you're convinced that those poor fellows who unwittingly consumed that tainted water will no doubt fall ill.

"No need to switch over to beer, the water's fine," your apparently less concerned neighbor says. "Relax. Limits are here to be overstepped."

"But it's a *pesticide* they found," you say, "that says it all. And we're talking about a threshold value for a bad chemical in our environment. It's a red line."

Your friend shakes his head. "Forget it. Those limits are set arbitrarily, they can be changed anytime. Like the debt ceiling in politics. When the money's running out, they just raise the debt ceiling."

Who's right? As ever so often, there is a lot of misinformation, but also a little bit of truth in both stances.

How Much Is Considered Safe?

We have learned in previous chapters that "safety" is a relative term, and that nothing is 100% safe. Nevertheless, the regulatory agencies keep coming up with tolerance limits for chemicals in the water, air, and our foods—individual threshold values below which exposure for humans should be safe, or, to use a more appropriate term, below which the human health risk is so low that it becomes "acceptable." How are the regulatory authorities able to do that? How do they know where exactly to draw the line in the sand?

The numbers are not made up out of thin air—they are based on rational calculations.

DOI: 10.1201/9781003346661-19 **159**

We do not a priori know where these safety limits are. For new drugs, pesticides, household chemicals, or toxicants in the food we may have some broad ideas, ballpark notions rather than precise figures, derived from comparison with similar compounds. However, history has taught us that we can make grave mistakes by assuming instead of testing.

Here is how it's currently done.

Because detailed epidemiological data in large human populations are not available for newly introduced chemicals for obvious reasons, and because *in vitro* data are unreliable for this purpose, the researchers initially get some information from animal studies. The chemical to be appraised is chronically administered to rats or mice, at different dose levels, usually in the diet or drinking water, for days, weeks, or months (for details on the safety assessment go back to Chapter 15). At the end of each treatment period, the readouts allow for establishing a dose-response curve. Next, a threshold dose must be defined—that maximal dose level at which a specified toxic response is not observed but above which a toxic response would be detected. That critical dose level is called the *no-observed-adverse-effect level (NOAEL)*. The NOAEL usually is expressed on a relative body weight basis (e.g., mg of chemical per kg body weight per day), and of course can vary depending on the specific toxicity endpoint chosen (e.g., neurotoxicity, liver function, and hormonal changes).

(For sticklers for accuracy, taking the chances that you'll accuse me of going off on a tangent again, I need to clarify something: there is no such thing as a zero-effect dose; everything has an effect, but it might be so minuscule that we cannot see it and therefore it becomes irrelevant.)

Next, the NOAEL from animal studies must be "translated" (extrapolated) to the human situation, which is quite tricky— remember, humans are not just big rats. At this point, this extrapolation can only be a guesstimate, at best. Not only are there differences between rats and humans, but also among humans themselves, as individuals might react to a chemical differently. There are a lot of uncertainties and what-ifs. Therefore, to be on the safe side (pun intended), "safety factors" (better:uncertainty factors) must be worked into the calculation. To do so, the rat NOAEL usually is divided by a factor of 10 for the translation from rats to humans, and then divided by another factor of 10 to account for the variability among different

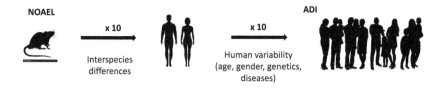

NOAEL

x 10

Interspecies
differences

ADI

x 10

Human variability
(age, gender, genetics,
diseases)

Figure 16.1 Calculation of a reference dose or acceptable daily intake for humans based on the no-observed-adverse-effect level in animals.

humans (children versus adults, elderly people, genetic differences, underlying disease, etc.), so that the total safety factor is 100. As you understand now, this projected no-effect level for people is an *arbitrary* number. However, even though being merely a rule of thumb, such an extrapolation has turned out to be judicious (see Figure 16.1).

Okay, so this putative no-observable-effect level can be taken as an acceptable maximal exposure level, a threshold below which no toxicity is to be expected. It is nothing else than the *Reference Dose (RfD)* that we've encountered earlier, or, for chemicals present in our foods, the *Acceptable Daily Intake (ADI)*, both basically meaning the same thing (to review the terms, turn back to Chapter 11). Again, both RfD and ADI are that daily amounts of a chemical that when ingested during an entire lifetime appears to be without any appreciable risk.

The 100-fold difference, i.e., the ratio of the NOAEL obtained from animal studies to the predicted or estimated safe human exposure or dose level is called the *Margin of Exposure (MOE)*.

Let's illustrate this with an example of a chemical we've discussed extensively in Chapter 10: the herbicide, glyphosate, a widely used weed killer.

The NOAEL for glyphosate, derived from a chronic, life-long toxicity study in rats was determined to be 100 mg/kg/day. Applying an MOE of 100, as mapped out above, we can now come up with a "safe" (acceptable) exposure level for humans of 1 mg/kg/day (100th of the rat dose, on a body weight basis). So, when the Food and Agriculture Organization (FAO) issued an acceptable maximal daily intake value for glyphosate of 1 mg/kg, you should understand how they came up with that number. Furthermore, you can appreciate now why a minor violation of the tolerance limit does not invariably entail toxicity.

(For regulatory toxicology devotees: the above-defined calculation of the generally accepted margin of exposure of 100 is not valid for those chemicals that are known to cause cancer by damaging the DNA, i.e., for genotoxic carcinogens. The reason is that, for those agents, a safe threshold dose is difficult to define because the shape of the dose-response curve at the very low end is uncertain. In those cases, an MOE of 10,000 or more is usually applied.)

Once the specific threshold level for a chemical has been defined, the question arises of how we can monitor that critical level in samples in our food, water, air, or consumer products. If it's a very small amount, do we have the tools to detect it, and reliably quantify it? You may be in for another surprise.

Detection Limits for Chemicals

In previous chapters, it has been repeatedly mentioned that the modern methods for detecting chemicals have become so sophisticated that it is possible to find residual amounts of chemicals almost anywhere, even when those quantities are too small to be of toxicological relevance. Without breaking the mold of the purpose of this book, it's time to add a few words about those methods.

The most frequently used method to detect, identify, and quantify residues of chemicals in samples like water, soil, or food extracts is LC-MS/MS (for connoisseurs: liquid chromatography coupled with tandem mass spectrometry). No need to go into detail—the method is fast, accurate, sensitive (i.e., it can pick up small quantities against the background "noise"), and specific (i.e., there's no interference from other compounds). Accurate measurements, therefore, are possible with small sample sizes. The method is incredibly sensitive; the limit of reliable quantification for a compound typically is in the lower ppb range, i.e., a few nanogram per milliliter. The limit of detection is even smaller; typically, in the higher ppt range, i.e., picogram per milliliter range. (As a reminder, 1 nanogram (ng) is one trillionth of a kilogram—see Appendix 1.) This is boggling.

Tainted Water

We have learned in Chapter 11 how to set the tolerated upper level of an unwanted chemical in foods, also called the *maximum residue limit (MRL)*. The same principle can be applied to contaminants in our water. This allows us to calculate the *maximum contaminant level (MCL)*.

Harking back to our paradigm, the pesticide glyphosate, we should now be able to better appreciate that the puzzling limits we're reading about are not made up of thin air. For instance, in the United States, the MCL for glyphosate in drinking water has been set to 0.7 mg/L. Assuming a person with an average weight of 70 kg drinks 2 L of water per day, the amount of glyphosate ingested via water (1.4 mg) would still be way below the RfD value of 1 mg/kg body weight (70 mg), leaving quite some leeway.

The situation can differ when we are talking about chemicals that might have a negative impact on entire ecosystems (aquatic life, birds, etc.). In those cases, the concentration limits can be much lower than expected from the calculations explained above. To stick with our example, glyphosate, considering its potential long-term environmental toxicity, the current MCL in all rivers, creeks, and lakes in Switzerland, was set to a level of 0.0001 mg/L.

Needless to say, those tolerance levels including the MCL can only be set, monitored, and enforced if the numbers are realistic in the broadest sense. For example, for unavoidable contaminants, the costs and feasibility of compliance must also be considered; if the efforts to bring down the MCL by a few additional points were fancifully high or simply impossible to implement, the limits likely are being reevaluated.

Looking back, your friend from the introduction to this chapter is correct in that the calculated acceptable or tolerable limits, although practical to work with, are not an all-or-nothing value, nor a tipping point for disaster. Given the 100-fold or higher uncertainty factor included in the calculation of the RfD or ADI, a 10% overstepping of the tolerance limit still leaves enough room to buffer the impact. A minor transgression of the tolerance limit should be taken seriously, prompting action, but it is certainly not a reason to panic. It doesn't catapult you out of absolute safety to full calamity. If anything, it means a minor increase in risk.

You will sometimes read in statements from governmental regulatory agencies that tolerance limits were calculated *based on all known facts at the time.* This may sound wordy and nit-picky (of course! based on what else?), but the reason is that over time the tolerance limits have been and will be shifted.

Here's a recent example.

Recalibrating the Tolerance Limits

We constantly acquire new knowledge over time. Scientific research is dynamic. Therefore, the current view of how we rate the risk of a chemical is not cast in stone. Neither are the tolerance limits for chemicals, defined at a given point in time. So, we shouldn't have been too surprised when the US Environmental Protection Agency in January 2022 added a new chemical, *1-bromopropane*, to the growing Hazardous Air Pollutants list. A new unsafe substance we're exposed to, out of nowhere? Why?

For starters, some background info. 1-Bromopropane was originally used in the production of pesticides, pharmaceuticals, and other products, but the real breakthrough came when it was introduced to replace other agents that had become notorious for their ozone-depleting properties. 1-Bromopropane is a volatile solvent increasingly produced for adhesives, cleaners, and degreasing chemicals (and previously used for dry cleaning). The good news for the environment is that 1-bromopropane does not belong to the notorious group of persistent chemicals; it evaporates into the air where it is broken down quickly. Also, it does not accumulate in food chains.

The initial enthusiasm, however, was dampened when it became clear that 1-bromopropane increased the risk for neurotoxicity in humans, mostly in occupational settings. After long-term exposure to 1-bromopropane via inhalation, workers developed symptoms including confusion, dizziness, or slurred speech. Animal studies revealed that 1-bromopropane was a rodent carcinogen, but there is no evidence so far that the chemical is a human carcinogen. New biochemical studies revealed the exact mechanism of how 1-bromopropane can attack and bind to certain sites in proteins. New data, time to act.

Back in 2003, a professional association of industrial hygienists in the United States had set a safety threshold for daily 8-hour exposure to 1-bromopropane at 10 ppm (i.e., 10 µL per L ambient air).

But some ten years later, after having learned much more about the risk associated with the chemical, the threshold was lowered by a factor of 100, now being set at 0.1 ppm.

Bottom line is, we shouldn't be surprised when the thresholds and tolerance limits change over time, nor should we blame the authorities for not knowing what they're doing when we read that the safety limits for a certain chemical are being revised downwards or upwards. We are just learning new stuff all the time.

Another discombobulating fact for consumers is that the exposure limits can vary from country to country, and, in the United States, even from state to state. For example, the exposure limit for a certain contaminating agent in the drinking water set by the US EPA may differ from that set by California by a factor of 10. What would it mean, and what are the reasons for such discrepancies? One possible reason is that the authorities may not always have considered the same endpoint for the assessment, e.g., have based their calculation on cancer risk as opposed to a non-cancer risk. In addition, different methodologies may have been used, or different uncertainty factors, or differences in the weighting of the risks. So, if you stare at a specific published exposure limit of, say, 70 ng/L per day, and you wonder why that number is not the same as in the area where you live, you should always see that number in its context.

But where should we draw a red line when we don't have enough data about a chemical, or, worse even, no data at all, without fudging a number? This will be the topic of the next section.

Setting Exposure Limits for Data-Poor Chemicals

Not every chemical we encounter in our food, water, air, or soil has been thoroughly tested for its potential to cause harm—there are simply too many substances around, and not enough resources available to do the job. Since there is no way all of those "data-poor chemicals" can be tested, we need to prioritize them. But how should one determine, on a scientific basis, which ones of these blank-slate chemicals need a detailed study and which ones likely present no appreciable health risk in the first place? The pragmatic concept is based on *experience* (uh-oh). Relax. It's still science.

To start with, we know from experience that most chemicals can be classified based on their chemical structure and the type of toxicity they exert, so we can roughly estimate their hazard. Next, we can make an educated guess about the expected exposure (dose, frequency, duration, and route of exposure). By combining these two key variables, the scientists can then define a so-called *threshold of toxicological concern (TTC)*.

This practical approach has initially been applied for chemicals in food (exposure via ingestion) but subsequently has been expanded to include chemicals to which we are exposed through inhalation or skin contact. (An exception is chemicals that bioaccumulate; there are limitations to this application.)

With the available information both about the chemical properties and the exposure, the different chemicals can now be assigned to groups of gradually ascending TTCs. For example, the lowest exposure threshold (0.15 μg/person per day) is applied for chemicals that react with DNA (potential carcinogens); at the other end of the scale, the highest exposure threshold (1800 μg/person per day) is applied to chemicals that are least concerning.

Again, this is a practical working tool to assist in prioritizing chemicals for further safety studies. As mentioned above, and similar to the traditionally estimated threshold values, the TTCs will be constantly refined as the scientific knowledge about the different chemical groups increases.

What does all this mean in practice? It means that when a certain chemical (say, a food additive present at a low level) has never been subjected to a rigorous full-scale safety testing procedure, we're not completely in the dark but still, have a ballpark idea about the threshold below which the risk for harmful effects is minimal.

TAKE-HOME MESSAGE

- The acceptable daily intake (ADI) or the reference dose (RfD) of a chemical is based on the no-observed-adverse-effect level (NOAEL) derived from animal studies, calculated with an "uncertainty factor" of 100 up to 10,000.

- Based on the ADI or RfD, acceptable (or tolerable) levels of a chemical in different foods or water can be calculated and expressed as maximum residue limit (MRL) or maximum contaminant level (MCL).
- The tolerance limits of a chemical are therefore arbitrary values and not set in stone; with newly acquired knowledge, they will constantly be reevaluated and adjusted if needed.
- Data-poor chemicals that have never been tested for toxicity can be broadly classified into groups allowing to assign them a threshold of toxicological concern (TTC), i.e., an exposure limit below which the human health risk is presumed minimal. This serves as a tool to prioritize the myriads of chemicals for further testing.
- *Lagniappe*: Experience is great but, as the saying goes, it enables you to quickly recognize a mistake when you make it again.

17
RISK ASSESSMENT

Risk Perception

Risk is the *chance* of something bad or harmful happening. In the context of this book, it is the chance of getting harmed by exposure to a chemical.

This simple definition implies a lot of uncertainty. How should one describe the possibility for something to happen without sounding wordy, iffy, and woolly? The consumers, though, want unambiguous answers—but, unfortunately, that's not always possible.

When someone says, "They've found pesticide residues on the veggies I'm eating every day, so I'm sure I'll get cancer," that's not a statement about risk, it's a yes-or-no view (or a belief, rather), unfounded, but also dead wrong because it's based on an erroneous inference. If we want to know more about those pesticide residues, the question is, how big is the actual human health risk? High, small, or negligible?

In addition, the term "risk" implies emotions. Therefore, the way how different people perceive risk is highly personal. Before we discuss the actual risk of chemicals on human health and the environment, and how to assess that risk, let's briefly consider individual *risk perception.*

We have touched elsewhere upon the different views of risk when talking about how that paramedic and that enthused biker each perceive the risk associated with being on the road. People often misjudge a risk when talking about a voluntary activity like, e.g., riding a motorcycle without a helmet. (Smoking and excessive drinking of alcohol is another example, but maybe not the best one because the "voluntary" part is trammeled by the addictive nature of those activities.) Another example of misjudging an unlikely event, in this case grossly overestimating it, is gambling in a casino. Although the odds against breaking the bank are astronomical, some people risk a fortune, and their rational thinking is clouded by the anticipation of the big win. But, to come back to our example, many people would perceive the risk of getting cancer from exposure to residual pesticides in our diet as high—again, gut feelings prevailing over actual facts.

DOI: 10.1201/9781003346661-20

Research has shown that cancer is among the most feared health hazard outcomes.

The key word here is *control*. When I'm doing something that I feel is totally under my control, like driving my speedy convertible, I'm willing to take a greater risk than when I feel I'm at the mercy of the pilot and that guy on the tarmac who last inspected the jet engine before takeoff. Part of the reason why people are afraid of flying is their impression that everything is out of their control. They block out the fact that the riskiest part of hopping on a plane likely is the drive to and from the airport.

In addition, something else can help to explain the deeply rooted fear of uncontrollable events: incomplete background knowledge. The more someone knows about a certain area or technology, the better they can grade the real risk. For example, a pilot has another risk perception of flying a plane than someone who experiences a white-knuckle bumpy flight for the first time. A geneticist has a different view of the risks associated with gene-modified organisms than a consumer. And, back to our pesticides, the more you understand about potency, exposure, and the other quantitative aspects of chemicals, the better you will be able to gauge the risk and the easier you can let go of unfounded fear—without losing perspective of the real dangers.

Deadly and Avoidable: Fugu

Here is an interesting example of individual risk perception and, based on that, *risk acceptance*.

Some brave people are perfectly willing to deliberately take chances, although they know that there might be an extremely high-potency chemical on their plate, on an exposure level high enough to potentially kill them. Again, one reason why they're doing that is perhaps the understanding that they have some sort of control over the situation.

I'm talking about tetrodotoxin, a natural toxin that is present in the pufferfish (called *fugu* in Japan). The toxin is confined to certain organs of the fish, like liver, ovaries, and skin, but is not found in the edible parts. Therefore, the preparation of the expensive fish dish takes high skills; in fact, a fugu chef needs a special license. Tetrodotoxin is one of the most potent neurotoxins in food, so if the fish is not properly prepared, there is enough toxin to kill a healthy adult (less than 1 mg is a deadly dose). Irrespective of the risk, each year some 100 poisonings from pufferfish are reported, about half of them are fatal. Talk about adventure.

(On a side note, for fans of natural toxins: you may ask yourself why the pufferfish carrying with them a hefty load of a potent neurotoxin do not get poisoned themselves but instead swim around happily in tropical waters. To understand that paradox, here is a quick detour: tetrodotoxin blocks certain molecular "channels" at the surface of our nerve cells, tiny pores in the membrane, which are of paramount importance for the nerve cell's function. The toxin fits snugly into that pore, plugging it from the outside so that the nerve is paralyzed, and the poor guy who has just ingested the poison is immobilized while being fully conscious, eventually dying from respiratory failure. Interestingly, the pufferfish's nerve cell channels are slightly different from that of other organisms due to a genetic mutation; tetrodotoxin doesn't fit into the tiny channels, leaving the pores open and functional and the fish itself unharmed.)

So, next, let's finally come to grips with how to assess in a rational way the risk of being exposed to all those chemicals around us.

Risk Assessment

Risk assessment is the process of *estimating the chance* of an untoward event to happen. If this sounds to you fraught with assumptions, you're right, but we have excellent tools to make the educated guesses as accurate as possible.

Let's translate this into the language of the toxicologist's world. To assess a certain risk associated with the presence of a chemical in our food, water, air, or environment, two questions must be answered: first, when is the risk *high* enough so that we need to take action, i.e., make regulatory decisions? And second, when is the risk *small* enough so that we can ignore it?

The stakes are high—but, as you can imagine, a reliable risk assessment is quite sophisticated because there are so many uncertainties involved.

For most chemicals, a risk assessment is done to gauge the *human health* risk (we'll talk about the environmental risk later). Typically, two slightly different approaches are used, depending on whether we want to assess the health risk associated with a cancer-causing chemical (a "cancer risk") or whether we're dealing with another type of toxicity (a "non-cancer risk").

Let's consider the *non-cancer risk* first because we should already be familiar with most of the metrics and terms. Remember the dose-response curve? For non-cancer endpoints (e.g., hormonal imbalance, liver injury, or neurotoxicity) it is assumed that there is a threshold in the dose response. This means that up to a certain dose of a chemical there is no detectable toxic response, which leads to the determination of a no-observed-adverse-effect level (NOAEL) in lab animals (see Chapter 16). As discussed earlier, when factoring in a couple of uncertainty factors for humans, we can now calculate an acceptable daily intake (ADI, for foods and drugs) or a reference dose (RfD, for environmental chemicals).

So, to proceed with our non-cancer risk assessment, all available data about the chemical in question must first be collected, including an assessment of its hazard. How "toxic" is it, and what is its potency with regard to certain endpoints? Next, the exposure will be assessed, estimating the daily dose, considering the different routes (dermal, inhalation, oral), and ascertaining if special populations might be at a higher risk (e.g., high exposure at the workplace). Third, with the chemical's toxicity profile now on the table, as well as the calculated ADI or RfD derived from the animal data, we can proceed to the final step. This is a characterization of the risk by dividing the expected dose (the estimated human exposure) by the ADI or RfD. Simply put, if the quotient is smaller than 1.0, no toxicity is to be expected; the risk is small or even negligible, and we're good for now. On the other hand, if the quotient is greater than 1.0, toxicity is to be expected, and the risk increases. All this is summarized in Figure 17.1.

(For toxicology buffs only—otherwise you may want to skip this paragraph: in many non-US countries, the regulations are different, and the unfamiliar terms and abbreviations can be confusing for outsiders, if not appalling. For example, in the EU, the so-called *REACH* regulation is the current overall framework for a risk assessment for both humans and the environment; the acronym stands for *R*egistration, *E*valuation, *A*uthorisation and Restriction of *Ch*emicals. Instead of the term RfD, under REACH regulations the exposure

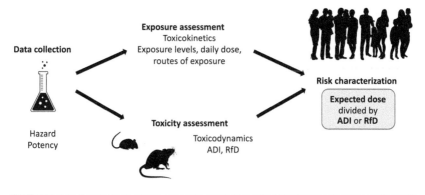

Figure 17.1 Non-cancer risk assessment.

level above which humans should not be exposed is called the *Derived-No-Effect Level (DNEL)*, which is basically the NOAEL again multiplied by uncertainty factors.)

Too notional? This is probably the most theoretical chapter in the book, so just bear with me for a bit.

When estimating *cancer risk*, we must recall (Chapter 7) that there is no "safe" level of exposure to a genotoxic carcinogenic substance, at least not theoretically (in practice, things may be different). Conservative views still consider the dose-response curve for cancer-causing chemicals as having no threshold dose. That means, even the smallest dose could elicit cancer—although we might not be able to detect it. That poses a problem for the regulatory agencies: one cannot determine a NOAEL in animals, and therefore not come up with an ADI or an RfD for humans (or a DNEL, see above).

Some, but not all, of the more current approaches take an alternative view. It seems as though at least some carcinogenic chemicals do have a dose range without a significant effect. For example, compounds that do not directly damage the DNA (tumor promoters) may exhibit such a "threshold." Let's not forget that the molecular defense shields and natural repair mechanisms (discussed in Chapter 8) afford protection against cancer-causing chemicals as well because they can fight off and reduce the effects of small doses of such chemicals.

Before folks versed in statistics start facepalming, here's a quick explanatory side note. Setting a "safe" dose range for carcinogenic compounds based on animal studies is difficult, to say the least because we are talking about *rare events*. In a typical cancer study (a so-called "rodent cancer bioassay" agreed upon by international guidelines), there are only three dose groups: low, medium, and high doses, plus control. The probability to find an animal with a chemical-induced tumor in the low-dose group depends on the group size; the more animals are exposed to the same dose of a chemical, the higher the chance a tumor would become manifest. However, the group size cannot be indefinitely increased for both practical and ethical reasons, nor can more dose groups be added to the study protocol. In other words, the "safe dose" is a moving target, a statistical problem.

To get out of this predicament, we must *extrapolate* (extend the trend by assumption) from the known high-dose effects to the unknown low-dose effects. Using all available data points, the specialists can calculate the steepness (slope) of the dose-response curve at the low-dose end and come up with a so-called *cancer slope factor (SF)*. (It is beyond the scope of this book to explain how exactly the factor is derived). Basically, the greater the SF value, the higher is the potency of the chemical to cause cancer. Thus, for a cancer risk characterization, the expected dose of a specific chemical is multiplied by the SF to give a *quantitative* estimation of the risk (see Figure 17.2.). Broadly speaking, the higher the resulting number, the greater the risk.

Importantly, the obtained figures (product of Dose x SF) describing the cancer risk for different chemicals are *relative numbers*. They

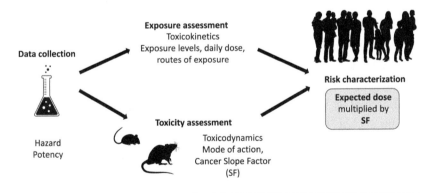

Figure 17.2 Cancer risk assessment.

Note: The *mode of action (MOA)* is a general term, referring to the major key events of a toxic response. For example, a genotoxic (DNA-damaging) MOA will be set apart from a non-genotoxic MOA.

provide an estimate of how many additional cancers on top of the ones that occur by other causes would be induced by a specific chemical in the general population over a lifetime exposure. Typically, the value is expressed as additional cases per one million people.

If you look up now and say, "hey, wait a minute," I can understand your growing concern. Probably you're asking yourself the same tough question that the regulatory agencies must answer: how small must the risk be that it is considered "acceptable"? Well, here is the number: it is generally agreed upon by international governmental consensus that 1 (one) additional cancer case in a population of 1,000,000 (one million) is acceptable (although I am not aware of any example).

I understand why people are getting upset, faces going red. Acceptable? Cancer is *never* acceptable. Every case is one case too many. Why don't the regulatory agencies finally do something to make it a safer world? People want *zero* risk.

Truth is, there is no such thing as zero risk, for anything, not just for chemicals. If the exposure is very low, the risk gets to a point where it becomes negligible because that case would disappear in the background (the "noise"). Don't forget, over a lifetime, *one out of three humans will get cancer*, for different reasons, some of which we do not understand. For a population of one million people, that is approximately 330,000 individuals who will get one or another form of cancer during their lifetime. If you add that additional cancer risk induced by a chemical, the number is 330,001 cancers. In real life, when a specific cancer is diagnosed in a patient, it is not possible to retrospectively identify the causative agent (e.g., make a pesticide responsible for it). All these numbers deal with average risks, and refer to human populations; if and where and when and to whom the real thing might happen is unknown.

To preempt any misunderstandings, and to admit that there are exceptions, there *are* certain cases of cancer that are so typical for a specific chemical, and at the same time extremely rare in non-exposed people, that one can venture to make a careful causal association. An example is a certain form of lung cancer called mesothelioma that typically arises in individuals exposed to asbestos. Another example is a type of liver cancer called angiosarcoma that typically occurs in people exposed to vinyl chloride, the chemical used for the synthesis of the plastic, PVC (polyvinyl chloride). Along the same rationale, when workers are exposed at the workplace to a specific chemical and a large proportion of them develop some type of cancer, a causal relationship can be established as well.

So far, we've been discussing the human health risk assessment. For other types of risks, e.g., for pesticides, the *ecological risk* is equally important, i.e., the risk a chemical may pose to aquatic and terrestrial animals, microorganisms, plants, and entire ecosystems. The approach is basically the same as for assessing the human health risk, but there are even more uncertainty factors because there are less hard data available and multiple species involved.

Rodent Cancer Bioassays—How Predictive Are They?

Some chemicals that can cause cancer in humans as evidenced by epidemiologic studies were previously identified as carcinogens in the rodent two-year cancer study (see Chapter 15). Among these are the notorious carcinogens like formaldehyde, vinyl chloride, TCDD ("dioxin"), radon gas, or asbestos. This might suggest that we have a perfect tool (albeit a complex and expensive one, and one that requires sacrificing numerous lab animals) allowing us to recognize and possibly eliminate new and dangerous cancer-causing chemicals.

Unfortunately, that's not the case. The positive correlation in *some* cases does not imply that we have been able to predict *all* carcinogenic compounds by this method. In fact, estimating the human cancer risk from chemicals is inherently more difficult than assessing the risk for acute, non-cancer toxicity. The reasons are manyfold—but a major reason is that rats and mice are not necessarily valid proxies for humans. It is difficult, if not impossible, to make predictions on cancer risk assessments when there are no human data available to confirm the validity of such predictive tests.

The current testing protocols for animal studies (rodent cancer bioassays) are based on many default assumptions. Among these is the false premise that chronic (two-year), high-dose exposure in rats and mice allows easy translation to a realistic low-dose exposure in humans. That high dose usually is the maximum tolerated dose (MTD, the highest dose that does not cause severe toxicity or mortality, determined from a series of preliminary studies). Another assumption is that the metabolism of a chemical at a high dose in rodents is similar to that at a low dose in humans. However, we know from *in-vitro* studies that there can be major differences in the metabolism of chemicals among species and across different dose levels. Finally, the uncertainty about the dose-response curve at the low end (as discussed above) adds to the list of conjectures.

The past decades have taught us that a positive cancer bioassay in mice or rats is not a reliable predictor of what might be happening in humans. In fact, the rodent bioassay has led to a gross overestimation of cancer risk—about half of all chemicals tested in mice or rats caused tumors at the highest dose (irrespective of whether the compound was natural or human-made). Especially liver and kidney tumors are frequently induced in mice but are rare in humans; on the other hand, prostate, pancreatic, colon, or cervix cancer are relatively frequent in humans but rare in rodents. Even among the two rodent species (mice and rats), there have been great differences in both the number of tumors and their site. An example is an environmental pollutant, naphthalene (keep reading for more on this).

So, the study of chemicals and cancer has taught us that humans are not just big rats. And mice are not little rats. We're not just talking about anatomical differences (e.g., in the airways), but we're also talking about differences in the metabolism of chemicals, immune surveillance of suspicious cells, or repair mechanisms of the damaged DNA that are critical.

A major factor contributing to the uncertainty in rodent cancer bioassays is the required high dose (MTD, see above). That dose is so high that in many cases it causes tissue injury in the animals. We know that repeated tissue/organ injury, followed by repair and continuous compensatory replacement of cells increases the background mutations and thus the chance that a tumor may arise. Also, chronic injury and inflammation generate oxidant stress (you may want to turn back to Chapter 8), which can further damage DNA. This may be irrelevant for the real-life low-dose exposure in humans.

A number of scientists, therefore, have claimed that the time has come to replace the traditional two-year rodent bioassays. But the question is, by what? Currently, new testing models are being developed and validated—models that take into account the mode of action (threshold or no threshold), incorporating toxicokinetics, and even making use of genetically engineered rodent models, enabling the researchers to recognize a cancer-causing potential at much lower exposure. However, it will take quite a while until new protocols will be fully validated, approved, and internationally implemented by the regulatory agencies.

Since these considerations have been quite theoretical—let's briefly consider an off-the-cuff example to illustrate the difficulty of extrapolating from animals to humans when it comes to estimating cancer risk.

Does Carpet Cleaning Cause Cancer?

A recent media scoop raises the question of whether carpet and upholstery cleaners are toxic, and might even cause cancer. Another alarming headline that may ratchet up the anxiety levels of many consumers, promote their general chemophobia and confirm their beliefs that the scientists have no clue, and the regulatory agencies don't do their job. As often the case, there is no mention of any dose or concentration, but dire warnings that those "toxic fumes" may cause cancer.

Better keep our hands off those rug-cleaning agents then? As one smart consumer said in a reader's comment: "See? That's why I've been cleaning only with chemical-free products from the eco store." Really? Frankly, I don't know how on earth that would've been possible, but good luck anyway with the removal of those stains.

The chemical of concern is *naphthalene*, one of the myriads of PAHs (see Chapter 11) that are present in fossil fuels and that arise during the combustion of organic matter. Naphthalene is also widely used in the manufacture of a variety of industrial products including dyes, surfactants, solvents, and resins. Other sources are open fires, traffic exhausts, or cigarette smoking; in fact, the compound is ubiquitously present. Because naphthalene, a solid white substance, readily evaporates at room temperature, exposure occurs mostly through inhalation. It's also been around for a long time as a pesticide to fight off insects like carpet moths that feast on natural fibers (wool)—you may be familiar with the typical smell from its use in old-fashioned mothballs. But how dangerous is it to inhale naphthalene? Here are the facts.

Naphthalene is a rodent carcinogen (relax—no reason to panic. Keep reading). Specifically, it has been associated, in those two-year rodent bioassays discussed above, with an increased incidence of benign tumors and a rare form of a malignant tumor (cancer) in the nose of both male and female rats, as well as benign tumors in the lungs of female mice, but not males. So, we have three different kinds of tumors, at different locations in the body, with distinct differences between the species, even closely related species (rats and mice). In humans, there is only sparse, anecdotal evidence for a possible association between naphthalene and cancer—not enough data to make a clear judgment.

What should be done with the available data? To create some clarification to the conundrum, a series of sophisticated studies by John Morris and his team at the University of Connecticut, as well as other groups, have revealed that the high doses of naphthalene used in animal studies were so toxic that they caused cell death and tissue injury in the airways. This does not happen at low, non-toxic doses, implying that there might be a threshold. Based on many sound scientific data, the IARC has classified naphthalene as a "*possible* human carcinogen," meaning that there is sufficient evidence in lab rodents but inadequate evidence in humans.

Again, this allegedly woolly statement does not mean the regulatory agencies are ducking the responsibility of a clear decision; rather, it reflects a common dilemma. The typical high-dose, lifelong exposure of rodents to a chemical that induces cell death cannot be directly compared to the low-level exposure of that chemical in our ambient air. (For those sedulous readers greedy for numbers: typical naphthalene air concentrations in cities are approximately 1 $\mu g/m^3$ (0.18 ppb), indoors concentrations some ten-fold higher. The US EPA has set the *RfC* (reference concentration) for nasal effects at 3 $\mu g/m^3$. For comparison, the "high dose" of naphthalene in the rodent bioassays (rat) was 314 mg/m^3 air (60 ppm), i.e., more than 100,000 times higher than the RfC.) We simply don't have enough human data available to make a firmer conclusion, but the general scientific consensus is that the low environmental naphthalene levels (at concentrations that do not induce cell killing) do not induce tumors at predictable rates.

So, while cleaning a dirty carpet may provoke dizziness, nausea, or loss of appetite (for various reasons), you don't have to freak out about inhaling a potentially life-threatening whiff of naphthalene. Extremely high, day-after-day tissue-damaging dose levels as used in the rodent bioassay cannot be directly compared to the low levels you've been inhaling.

Risk Management

Risk management is the *control of risk by eliminating the conditions that contribute to that risk*. The goal is to minimize the risk or avoid the impact of hazardous chemicals and ultimately protect the consumers, the human populations, and the environment.

Risk management is also about decision-making. Regulatory toxicologists are responsible for deciding whether a specific chemical (e.g., a new drug) poses a sufficiently low risk to be marketed. The regulatory agencies also establish the standards for the amount (or concentrations) of a specific chemical that is permitted in the air or in drinking water. There are big geopolitical differences with regard to those guidelines, regulations, and laws, but increasing efforts have been made to harmonize them at an international level. As mentioned, in the United States, the Environmental Protection Agency (EPA) is responsible for regulating pesticides, whereas the US Food and Drug Administration (FDA) regulates foods, drugs, and cosmetics. Clearly, it is beyond the scope of this book to expand on international regulatory toxicology.

However, risk management can also start on the individual level, e.g., by removing potential sources of exposure to chemicals. For example, this can be done by donning appropriate protective gear, goggles, gloves, etc., when handling hazardous chemicals. For foods, this can be done by washing produce prior to consumption and aiming for a balanced, nutritional diet. Many of the risks can be avoided altogether, such as by not smoking, reducing alcohol consumption, or avoiding recreational drugs.

The art is to distinguish between those chemical risks that pose a major challenge and those that bear a small or negligible risk. We should focus our efforts on the former and not waste too much time and energy on the latter. For example, in Chapter 13, we examined a bunch of chemicals that pose a significant human health risk and explored what can be done to reduce or entirely avoid the exposure. Unfortunately, for the average consumer who doesn't have scientific first-hand information or doesn't know how to interpret the scientific lingo, it is quite difficult to make the distinction between real risks and reputed risks, unless someone communicates the facts.

As we will see in the next chapter, talking about the risks can sometimes be a bit, well, risky.

TAKE-HOME MESSAGE

- How we perceive risk is personal and not always rational. We are prepared to take greater risks when we think we have control over the situation.
- Human *non-cancer risk* assessment is done by estimating the human exposure and the RfD (or ADI), assuming that there is a safe threshold dose. The risk is characterized by dividing the expected dose by the RfD (or ADI).
- Human *cancer risk* assessment is done by estimating the human exposure and calculating the cancer slope factor (SF), assuming that there is no safe dose. The risk is characterized by multiplying the expected dose with the SF.
- One out of three people will develop cancer over their lifetime. In view of this high incidence, an additional cancer risk of 1 per 1,000,000 people has been agreed to be "acceptable" for low-level exposure to contaminants. However, none of these additional cases can be measured or identified.
- For pesticide residues, the human cancer risk is very small.
- The two-year rodent bioassay to detect chemicals that may cause tumors is not a reliable tool for predicting cancer in humans. High doses that damage cells, leading to compensatory regeneration of tissues, cannot be directly extrapolated to low environmental doses.
- Regulatory agencies decide whether the risk is sufficiently low for a new product to be marketed, and they establish the quantitative standards for chemicals in the environment.
- *Apropos*: Zero risk is neither realistic nor possible.

18

GAUGING THE RISK AGAINST THE BENEFIT

The Benefit Must Outweigh the Risk

An important part of the risk management process (regulatory decision-making) involves an assessment of the risk-benefit ratio. This holds true for drugs, consumer products, pesticides, as well as other industrial chemicals. Let's not forget that whenever we accept a certain risk (including the human health risk from exposure to chemicals), it's because we're expecting something beneficial in return.

Careful benefit-risk considerations are an integral part of the regulatory review of marketing applications for new drugs. For example, certain chemotherapeutic agents may be highly toxic, killing not only tumor cells but also healthy cells, even cause DNA damage, and yet they have been approved to treat people who have cancer. The key is, ultimately the benefit should outweigh the risk.

After all that dry theory about risk assessment, here are two nuts-and-bolts applications of such risk-versus-benefit considerations.

Healthy Eating

The suspected presence of pesticide residues on fruits and vegetables has raised concerns about the safety of consumers. Therefore, many risk assessments for these chemicals have been conducted worldwide. We have learned that, yes, they are present, multiple pesticides in and on your foods, albeit in tiny amounts. Best to cross those items off your shopping list?

A research team around Mathieu Valcke at the National Public Health Institute and University of Montreal, Canada, has done an interesting nutrition study to weigh the human health risk versus the benefit of eating fruit and vegetables. They analyzed a huge amount of data on more than one hundred different pesticides found on the 30 most consumed fruits and 30 most consumed vegetables over many years. They computed the individual daily exposure to pesticide residues in eastern Canada and estimated the excess tumor risk by applying

DOI: 10.1201/9781003346661-21

the known cancer slope factors. They came up with a total lifetime cancer risk estimate of 0.00033 (i.e., 3 extra cases per 10,000 persons exposed through life)—a number that is not negligible.

Now, in the same study, they estimated how many cancers were likely *prevented* by eating healthy fruits and vegetables. The number that emerged was approximately 330 prevented cases per 10,000 persons throughout life. So, with all the uncertainties involved, and to put it simple, the benefit outweighed the risk by a factor of one hundred. (Just as a reminder—it is not possible in a population to identify a person who has cancer that might have been caused by pesticide residues, much the same as it is not possible to know who did not get cancer because they had consumed fruit and vegetables all their life.)

Another research team in the United States did a similar study and figured an even greater number (approximately 2,000) for the benefit-over-health risk ratio. So, consumers should keep that in mind when eating fruits and vegetables (Figure 18.1). At the same time, though, the chemical residues must be continually monitored by the authorities and action taken if needed.

A similar example highlights the risk-versus-benefit consideration of consuming salmon. Salmon, as other large fish, on the one hand, contain residues of toxic contaminants (mostly methyl mercury, or polychlorinated biphenyls that are persistent in the environment—you may recall those chemicals from Chapter 13) that, depending on the exposure, may have adverse effects on human health. On the other hand, salmon contains a lot of those famous omega-3 fatty acids (the "heart-healthy fats"). So, is it okay to indulge in salmon or should the fish rather be avoided? Calculations of the overall risk-benefit ratio are complex, but Christopher Rembold at the University of Virginia tried to

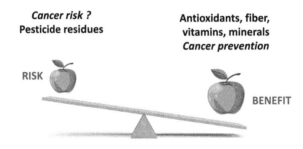

Figure 18.1   Balancing risk against benefit.

quantify this anyway. He contrasted estimates done by the US EPA of the increased cancer risk from contaminated salmon with a clinical trial in Italy that had demonstrated reduced mortality in coronary artery disease patients after daily ingestion of omega-3 fatty acids from fish. The conclusion of his complex calculation was that some 100 Italians with coronary heart disease who eat approximately 1 kg salmon per month would avoid death for every Italian who develops cancer from those chemicals. Obviously, one cannot directly compare these two things, but it seems evident that eating salmon has clear health benefits which must be balanced against a rational concern about toxic residues.

A Day at the Beach

A variety of ultraviolet (UV) filters, aka sunscreens, have been developed to protect us from harmful exposure to sunlight (both UV-A and UV-B). However, over the past years, their widespread use has raised concerns about a possible human health risk, as well as an environmental risk for aquatic ecosystems. Here's the reason why: certain chemicals in those sunscreens have been found to be endocrine disruptors (see Chapter 7), featuring weak estrogenic activity in certain experimental models. Deeply unsettling reports, raising doubts about whether we should set the dermatologist's advice at naught, and stop to lavishly slather sunscreen on our skin as recommended. We all love outdoor activities—so what should we do?

Let's do a quick risk-benefit analysis.

The benefit of using UV filters is a no-brainer. They reduce or prevent sunburn and skin damage and, most importantly, diminish the risk of developing skin cancer.

How about the risk of UV filters for humans and the environment? This is a bit more complex. To simplify things, let's focus on the most widely used ingredient in UV filters, benzophenone-3 (BP-3).

The first potential risk to consider is a possible deficiency of vitamin D—recall that sunlight (UV-B) stimulates the production of vitamin D_3 in our body. If you shield away UV light from the skin by applying a sunblocker, you might decrease your vitamin D levels. However, we can tap into other sources of vitamin D precursors: through our diet. In fact, population-based prospective studies did not find any significant deficiencies of vitamin D caused by applying UV filters.

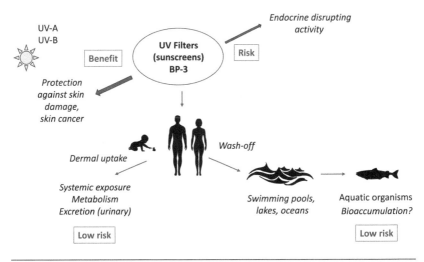

Figure 18.2   Risks and benefit of UV filters.

The other, more concerning potential risk is the alleged endocrine-disrupting potential of UV filters (summarized in Figure 18.2). You may have heard that in 2018, the State of Hawaii banned the sale of all sunscreens containing two specific organic UV filters (one of them is BP-3), with the aim to protect damage (bleaching) to coral reefs through the release of these chemicals into the seawater. There must be a reason for this.

BP-3 from sunscreen products is released into the environment mostly by recreational activities and is found in seawater, but also in freshwater and swimming pools. Concentrations are in the 100–500 ng/L range; higher levels have been found in "hot spot areas" and wastewater influents (up to 10 µg/L). Studies in zebrafish, a widely used test organism for aquatic toxicity, have revealed that developmental abnormalities and sexual maturation can occur at concentrations of approximately 400 µg/L and higher, which is a high concentration, approximately 1,000-fold higher than the average concentrations in ambient water. Thus, it is generally assumed that BP-3 poses a low risk to freshwater ecosystems. A problem could be the chemical's bioaccumulation potential; therefore, long-term studies will be needed.

As to the human health risk, one must again be careful when extrapolating data obtained *in vitro* to the *in vivo* situation. It was found that in cell cultures, BP-3 binds to the estrogen receptor and exerts weak estrogenic activation at high concentrations. In humans, uptake occurs primarily via the skin after topical application (people usually don't drink seawater). In fact, up to 10% of the applied dose can reach systemic circulation, and maximal plasma concentrations of BP-3 were 200 ng/mL, which is a relatively low concentration. In addition, BP-3 is rapidly metabolized in humans and excreted via urine. To date, no significant effects of BP-3 on the levels of reproductive hormones have been reported for humans. Therefore, BP-3 and similar compounds in sunscreens do not seem to be of toxicological concern; nevertheless, the situation warrants continued attention.

(Just to complete the picture, there is another type of sunscreens, the so-called "physical sunscreens," like titanium dioxide and zinc oxide. These chemicals are present as nanoparticles (see Chapter 20) that reflect and scatter the UV light. The small (20–100 nm) nanoparticles do not penetrate the skin and remain on the top layer of the skin.)

So, for UV filters, the risk-benefit analysis clearly is in favor of the benefit, which doesn't mean that we should be careless about future developments in the field. We can always switch to even less concerning alternatives once they become available (green toxicology). But for now, you may still enjoy your day at the beach.

TAKE-HOME MESSAGE

- The risk of a chemical must always be weighed against its benefit. Risk-versus-benefit assessments are a key part of risk management and decision-making.
- For fruit, vegetables, or fish, the health benefit greatly outweighs the risk from residual pesticides or persistent environmental chemicals. Nevertheless, contaminant levels should be lowered as much as possible.
- The benefit of UV filters to protect the skin is obvious; although BP-3 and other components of sunscreens exhibit weak estrogenic activity *in vitro*, the human health risk (disruption of hormone levels) is negligible.
- *Incidentally*: The old saying, "an apple a day keeps the doctor away" is still valid (but you might want to wash the apple first).

19

RISK COMMUNICATION

The Daily Dose of Hazardous Headlines, Toxic Information, and Risky Conclusions

Why framing a chemical as an environmental poison when in fact it isn't?

Scaremongering by the media? Ignorance of the reporters? Because it's an in-thing to bash the chemical industry? Or is it meant as a wake-up call for the viewers to help save the environment?

Maybe the simple reason is that the scientific facts are not readily available for everyone, and therefore poorly communicated, in bits and pieces, some incomplete, others unfounded or flat-out wrong.

Let me be clear: the vast majority of science journalists are serious and hard-working people who master the art of science communication and have done their homework. Most contributions are excellent, some outstanding, but, unfortunately, not all of them.

Let's illustrate this with an example.

In recent years, chemical analyses performed by trusted laboratories in several countries have revealed that a substance called *trifluoroacetic acid (TFA)* was present in the drinking water—not just in one sample, but in all samples collected at different locations across the country. Just imagine! The compound was even found in bottled mineral water—reason enough to point a finger at the producers. Trifluoroacetic acid is present in beer and tea, too (no cheap excuse anymore to switch to beer). Obviously, it doesn't belong there, so the media were on the spot, cranking out newspaper articles and TV shows for consumer protection, using high-pitched titles like "Environmental Poisons in Our Drinking Water."

An eye-catching headline like this will stir up the sleepiest citizen over their morning coffee and make them want to read more. Kudos to the reporters—this time they even mentioned the concentrations: on average, we're talking about 1–5 μg TFA per liter of drinking water or beverage. But does this number help the alerted reader gauge the risk? On top of that, the consumer learns, TFA is everywhere and anywhere on this planet and you can't escape the threat, is persistent in the environment and doesn't go away, and is toxic to aquatic life

DOI: 10.1201/9781003346661-22

(the wording perhaps implies that all fish might be swimming belly-up). Plus, certain pesticides are among the likely sources of TFA, of course. There we go again. The perfect disaster. Ample food for concern, a reason to blame the authorities for not doing enough, to accuse the manufacturing industry of slowly killing our environment. The rest of the day is ruined—and yet, the reader still doesn't have a clue what the heck TFA is.

But, hand on heart—do you honestly think anybody would read that article if the headline were, for instance, "Tiny Traces of Ubiquitous Chemical in Our Drinking Water that Poses a *De-Minimis* Risk to Human Health"?

So, let's dissect, piece by piece, what we know about TFA, then make a rational assessment.

What is TFA? Trifluoroacetic acid is a common degradation product of hundreds of thousands of different fluorinated organic chemicals widely used in industry, medicine, agriculture, and household. The end product after degradation is TFA, which is an extremely stable chemical that is resistant to any further degradation in the environment.

What are the sources of TFA? The major sources of TFA are the newer refrigerants, used to replace the old, ozone-depleting "bad" refrigerants. These refrigerants are volatile compounds that escape into the air where they are degraded by light. Other sources include a host of pesticides that contain the trifluoro- part, and certain pharmaceutical drugs.

Now here's an interesting point: besides coming from human-made sources, TFA is also of natural origin. It is found in all oceans at concentrations of approximately 200 ng/L water, even in remote locations, and the concentrations do not greatly vary with depth. Perhaps surprisingly, TFA has been found in deep (more than 1,000 m) water of the Atlantic. The sampled TFA was found to be older than 1,000 years, based on radiocarbon dating. The natural source is likely the deep hydrothermal vents, i.e., fissures in the sea bottom, as the measured TFA concentrations were higher there than in the surrounding waters. So, to put things into perspective: the TFA concentrations in drinking water (see above) were thus 5–25 times greater than the background concentrations in the oceans.

How does TFA get into our drinking water? Upon degradation of the fluorinated compounds in the atmosphere, TFA dissolves in water droplets and reaches the surface with rain or snow, and thus gets into the soil water, groundwater, springs, rivers, lakes, and ultimately the oceans. As mentioned, TFA is extremely stable; it will persist in the environment for decades, probably centuries. The good news, though, is that TFA does not bioaccumulate, unlike other "forever chemicals."

How toxic is TFA? (For toxicology fans: TFA is a strong acid and could, in its pure form, cause acute harm, like, e.g., hydrochloric acid. However, this is not relevant; in the environment, TFA is dissolved as a salt.) The concerns, if any, are about chronic rather than acute exposure. Mammals are not sensitive to TFA or its salts, even when exposed to it for a prolonged time. (For math freaks, let's do a quick calculation. As a rough estimate, the NOAEL of TFA administered daily to rats for 90 days is approximately 10 mg/kg body weight. Assuming humans drink on average 2 L water per day, containing 5 µg TFA/L, to assume the worst case, they would ingest 10 µg TFA per day. Using an average body weight of 70 kg, it would result in a daily dose of 0.14 µg TFA/kg body weight/day. Although rats and humans cannot be directly compared on a mg/kg body weight basis (you may recall the reason why from Chapter 15), it is obvious that this 70,000-fold difference is huge. Theoretically, a person would have to drink some 140,000 L of water each day to be in the ballpark range of the rat NOAEL.)

TFA has not been reported to cause mutations, to be carcinogenic, or to affect reproduction in mammals. As to aquatic organisms, toxicity can be induced, but only at exceedingly high concentrations; the NOAEC for Zebrafish or Daphnia (both are widely used freshwater animals in environmental research) is > 1,000 mg/L—more than one million times greater than the average TFA concentrations found in water.

What is the human health risk and environmental risk of TFA? Based on all available evidence to date, the risk for humans and terrestrial animals is minimal. The margin of exposure (MOE; flashing back to Chapter 16) is extremely large. Even if the exposure increased, the human health risk would still be negligible; the same holds true for plants and aquatic animals.

In a nutshell, TFA likely does not pose a risk for us, nor will it be a significant risk in the near future. Nevertheless, we should reduce its release into the water as soon as we have alternative products, and thus prevent any long-term risks before they even occur (stay tuned for the section on green chemistry in the next chapter).

Wouldn't that be a worthwhile topic for a newspaper article or TV show, comforting and informative, and, for a change, less depressing?

Balancing the Information

Unfortunately, the discussions about the risk of chemicals for both human health and the environment are mostly one-way communication. Most average consumers are flooded with unfiltered snippets of information from the Internet, newspapers, blogs, social media platforms, and TV programs. Sadly, the information often is not only incomplete, but at times unbalanced, or flat out wrong. Most people don't have the possibility to read the original research articles because they rarely have access to the databases that list the humongous (and growing) mountain of scientific information, and if they had, they likely would not understand all of the scientific terminology. By the same token, they have no access to national or international meetings where the newest scientific insider information is presented to the specialists in the field. So, from where should they get unbiased, rational information about the toxicity of specific chemicals?

Where is the go-between? Who is the facilitator?

Two kinds of people should accept that responsibility: first, those media reporters representing the "science and technology" sections, and second, the scientists themselves.

However, this doesn't always work in an optimal way. Reporters don't necessarily have a sound background in toxicology (they may have a general background in science though). Also, lurid headlines sell better than a soothing article. But to be fair, reporters often are under deadline pressure and simply don't have enough time to do all the homework they should do. So why not consult a specialist in the field? Same thing—a specialist with a measured and correct approach, trying to carefully balance the pros and cons of a point, is less of a crackerjack than someone with a pointed view and trenchant language who makes a strong case. A discussion among people with extreme views has higher audience ratings than a boring lecture by a scientist who will weigh twice every word they say. Incendiary headlines sell better than a soporific treatise.

So why are the scientists themselves not ready to become more active in public education? One of the reasons is that everybody is busy with teaching, keeping the lab running, doing research, and writing grant proposals and research papers for the top journals in the field. In addition, many research scientists often feel that there is hardly any incentive in exposing their views publicly—but there should be some time left to communicate the science in laypeople's terms, with a balanced view, building on the facts. It's time to bridge the gulf between the specialists and the non-scientists. Otherwise, the gap might even get wider, fostering misunderstandings and nurturing resistance against the rational science-based gain of knowledge.

But here's another thing: would the average consumer actually believe what the researchers tell them? Would the experts be able to convince people that, say, chemical X, which has fallen into disrepute, in reality, poses a small human health risk because the exposure is simply too low? Or, vice versa, that chemical Y, which has been widely used and generally assumed to be harmless, poses a higher risk because the latest research has found the hazard to be higher than previously thought.

Unfortunately, skepticism about science has sprouted in recent years. Furthermore, scientists in academia have often been accused of being biased because some of them receive financial support for their research from industry or governmental agencies. This false picture needs to get straightened. The overwhelming majority of scientists will stick to the facts and base their conclusions on hard data. True, in many instances, corporate researchers and scientists in academic institutions closely collaborate with each other for several plausible reasons. First, chemical companies do not always have the time and resources to do in-depth academic research—so they'll outsource parts of a project. Second, many academic scientists are dependent on private grants and funding from the industry—but they will publish their results. Their career would be crashed forever if they put a slant on their data. Finally, the days when a single researcher could make a groundbreaking discovery are long gone. In view of the growing environmental challenges (not only from a toxicology perspective), scientists from basic research, applied research, and regulatory agencies increasingly work together because this creates synergies and avoids duplication and waste of resources.

To be fair, though, the ever-increasing media reports on the health risk impact of "bad chemicals", while giving too much weight on the subject of hazard and neglecting the quantitative aspect (exposure and dose) may have a good side to it, too—they make the average consumer aware of a potential issue, hopefully stimulating them to get more reliable first-hand information, and becoming interested in the subject of toxicology.

TAKE-HOME MESSAGE

- Popular articles or TV shows covering the risk of chemicals in our environment, food, water, or air can be excellent and educational, but some are superficial, sometimes warped, neglecting to mention the quantitative aspect, thereby skewing the risk.
- Read/watch such reports with a critical mind and, if possible, read the original reports.
- The risk of TFA toxicity to humans is negligible.
- The scientists should come off their perch and communicate the risk in plain language.
- Scientists from universities, industry, and governmental agencies increasingly work together to create synergies and solve problems—listen to their voice.
- *As a mere fleabite*: If the facts don't fit one's view of the world, don't change the facts.

PART IV
THE FUTURE

20

TOXICOLOGICAL CHALLENGES

Chemical Medleys and Data-Poor Chemicals

For most examples we've talked about so far, we've always been focusing on one single chemical we are potentially exposed to, and we have learned how to make a reasonable risk assessment based on sound scientific data for that specific compound. The reality, however, is different. We may be immersed in a gazillion of different chemicals around us, and we can detect them individually and even quantify them, we get that—but how about entire groups of related chemicals belonging to the same chemical family, like "the dioxins" or "the PFAS" (stay tuned for more), or "the PAHs" (reeling back to Chapter 11), chemical cocktails washing around us? And we're not talking about just a couple of chemical family members showing up simultaneously—rather, we're dealing with hundreds or even thousands of closely related compounds inundating us as members of one chemical superfamily. The problem is, while these compounds may be similar in structure, they can vastly differ in terms of their toxicokinetic behavior, their potency, or the toxic response that they elicit, and their residual amounts in food can greatly vary. How on earth can we make an overall risk assessment for those compounds? Are we stranded? Back to square one?

The most logical thing would be to deal with each one of those chemical group members individually, one after the other, which would take forever—but there is another, even more serious problem: we are often confronted with a lack of dependable safety data for many of those compounds (therefore called "poor-data chemicals"). This holds particularly true for industrial products with low acute toxicity that are manufactured in relatively low amounts.

A typical example that combines both issues discussed above, i.e., a chemical mix of poorly characterized chemicals, is the *PFAS* (or, for toxicology mavens: the *perfluorinated alkyl substances*). PFAS are a large group of industrial chemicals comprising more than 4,000 related compounds (so-called congeners). Developed in the 1940s, they have been used for cosmetic and household products (e.g., Teflon), outdoor

DOI: 10.1201/9781003346661-24

clothing, food packaging materials, but also for use in fire extinguish-
ers (foams). Their popular use is based, among other characteristics,
on their poor reactivity, heat resistance, and stain and soil-repellant
properties.

The flip side of their chemical stability is that PFAS are not
degradable in the environment. They are therefore ubiquitously present and have been detected globally in the groundwater, lakes, oceans,
air, and in plants. They bioaccumulate and are found in animals and
humans. Two important ones (PFOS and PFOA) have recently been
banned on account of safety concerns, but a host of new ones have
been appearing.

Unfortunately, there is no solid toxicity database available for the
individual PFAS, which is understandable for a family of thousands
of different compounds and counting. For most of them, there is little
or no information on their potential human health effects. Hazard
characterization and risk assessment have generally been grounded on
a few typical exponents (e.g., the legacy chemical, PFOA), but that
may not represent the rest of them.

How about exposure? Except for occupational exposure, e.g., at
the workplace, direct exposure to PFAS from handling consumer
goods is a minor route. The major pathway is indirect exposure via
contaminated groundwater and foods from marine food chains
(see Figure 20.1). Let's recall, we're not talking about one or two
chemicals—we are exposed to a mélange of thousands of different
PFAS with very few toxicological data available, if at all. The only way

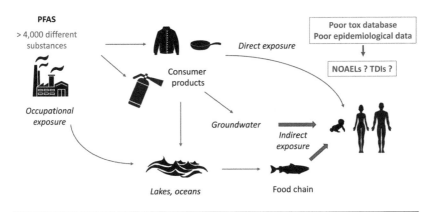

Figure 20.1   Environmental fate of PFAS.

out seems to rely on epidemiological studies, which not only takes a long time but also does not always allow for establishing a clear causal relationship (harking back to Chapter 9, a simple correlation is no guarantee for causation). For example, an association between exposure to PFAS and adverse immune outcomes in children has been observed, but the underlying mechanisms are unknown.

Different agencies in different countries have proposed a lifetime health advisory level for one of the notorious PFAS species, PFOA (which has been phased out, but which is still around in the environment) in the range of 1–10 ng/L drinking water. This is a low concentration, lower than the health risk threshold set by the EPA just a few years earlier, and due to new scientific data; in fact, the levels are so low that they are barely measurable (you may want to revisit Chapter 16 to recall the stunning detection methods). However, for the rest of the PFAS currently in use, there is probably a wide range of hypothetical risk levels. Thus, it is simply impossible to come up with a TDI (tolerable daily intake) for PFAS as a group.

So, it's difficult to define exposure limits because we know so little about these chemicals. What we do know, though, is that they persist in the environment. Which nicely segues into the next section.

Nanoparticles and Microplastics

The study of potential adverse effects of *nanoparticles* (nanotoxicology) is a relatively young area in toxicology. Nanomaterials (per definition, materials smaller than 100 nm) have been increasingly used in medicine, computer technology, household products, cosmetics, and many other industrial applications. The chemical composition of nanoparticles varies, but because of their small size, they have unique properties. While the technology itself has been soaring, research on the potential toxicity of nanoparticles is lagging behind.

The major route of exposure to nanoparticles is inhalation, but dermal exposure may also be a concern. In lab animals, small particles are usually deposited in the upper airways, but the even smaller nanoparticles can reach the deepest parts of the lungs, causing inflammation and damage. Importantly, they can enter the bloodstream and distribute throughout the body. In cell cultures, many nanoparticles can cause oxidant stress and induce cell death (apoptosis).

Although we have learned a lot from animal toxicity studies, we are still far away from being able to assess the human health risk of nanoparticles. In addition, some of the studies are controversial. Exposure limits will have to be defined for occupational exposure. To do that in a serious way, we need more toxicity studies before an international regulatory framework can be established.

Another still insufficiently studied area is the toxicity of *microplastics*, even though the problem is not new. Plastic (synthetic polymers) production has been increasing in recent years—some 400 million tons of plastic are currently being produced annually. A lot of it ends up in the environment, mostly stemming from textiles, tires, or city dust (it's not just the crumpled water bottles). Plastic materials are persistent in the environment; they degrade at an extremely slow pace to smaller particles, ending up as "microplastic," i.e., particles between 5 mm and 1 µm. Microplastics are ubiquitous in marine and freshwater, where they are ingested by organisms, even making their way into the food chain (or food web, to be more accurate).

Microplastics are found everywhere, but little is known about their human health risk. Exposure of humans to microplastics (by ingesting microplastic-containing foods, or by inhalation) currently is low, and there are no published studies yet on their potential effects on organisms. The first preliminary studies were performed in cell cultures; extrapolation to the *in vivo* situation is challenging. Another potential issue, besides being a possible direct toxic hazard, is that the tiny particles could be a vehicle for toxic metals or even pathogens. Let's not forget that due to the particles' tiny size, their total relative surface in relation to their mass is enormous.

Another challenge, distinct from that related to the larger particles, is the presence of ultrafine microplastic particles (smaller than 1 µm). Such "nanoplastics" can theoretically cross cell membranes and get inside a cell, causing an inflammatory response, similar to the effects of other nanoparticles, as discussed above. Indeed, a recent study by Marja Lamoree and her team at the University of Amsterdam, The Netherlands, has found tiny plastic particles (≥700 nm) in human blood from 22 volunteers (at a mean concentration of 1.6 µg/mL), demonstrating that plastic particles are bioavailable for uptake into the bloodstream. Whether this might pose a human health risk is currently not understood.

When will the scientists be ready to conduct a comprehensive human health risk assessment for microplastics? It is too early to gauge the risk on a rational, scientific basis. The first thing that needs to be done is to assess the exposure (which shouldn't surprise you). Next, the hazard must be determined. Both factors are not yet known at this time. We are just at the beginning. The only thing we can say is that plastics have been around for a long time, and if there were drastic effects on human health, we would have noted them. However, here is a sobering reminder: the particles are out there; even if we'd radically cut back on plastic production (which is unrealistic), we would still have to deal with the issue for many years to come.

Forever Chemicals

Persistent chemicals that stay in the environment for a very long time are called "forever" chemicals, although that's slightly exaggerated. We have already encountered some examples: certain long-lived pesticides like DDT (see Chapter 10), PFAS (see above), dioxins, and metals (see Chapter 13). Such chemicals can bioaccumulate in the body and, if they get into the food chain, undergo biomagnification. Importantly, longevity in the environment does not necessarily imply that they are high-hazard chemicals—e.g., plastics are chemically inert, yet accumulate over time, and the real risk may arise from the tiny degradation products (micro- and nanoparticles, see above). Some of these forever chemicals have been banned years ago but are still found in the environment; in fact, if, say, all PFAS were banned with immediate effect, they would remain in the soil and water for decades to come.

Green Toxicology

Everybody knows by now that in order to protect organisms from the potentially toxic effects of hazardous chemicals we need to reduce the exposure, at least try not to exceed the established limits for safe use. Another, maybe even more reasonable step in the same direction would be to *reduce the hazard* of human-made chemicals in the first place. But how can this be done—hazard being an inherent property of each chemical?

A relatively new area in the field is "green toxicology" (a domain of "green chemistry," aka sustainable chemistry). Green chemistry aims at designing chemicals and processes that reduce or eliminate the hazard. The point is to prevent adverse effects *before* they can happen, i.e., fixing the problem upstream, at the beginning of the chain (or network) of events, as opposed to dealing with the issues after they occur. It's better not to synthesize a hazardous chemical in the first place, let alone bring them into the world. Traditional methods have been focusing on the other end, i.e., sequestering hazardous compounds once they're out there, cleaning up spills, or fighting pollution.

Green toxicology is not something created by starry-eyed idealists; its application is a serious attempt to learn from the past. The increased use of "greener" chemicals will undoubtedly reduce the risk of adverse effects on humans, animals, and plants. It will also prevent that such newer, greener chemicals persist in the environment where they might bioaccumulate. Finally, the risk for manufacturers or users of such chemicals will be greatly diminished.

Meanwhile, and likely for a long time to come, we'll have to focus on the exposure and grapple with the risks posed by all the current, both natural and human-made, chemicals surrounding us.

TAKE-HOME MESSAGE

- It is not possible to make a sound, science-based risk assessment and define tolerance limits of mixtures of related, data-poor chemicals (e.g., PFAS).
- Nanoparticles and microplastics are ubiquitous, yet we do not yet have sufficient scientific data on their hazard, exposure, and risk to human health and the environment.
- Forever chemicals may hang around for years or decades because of slow degradation in the environment and poor metabolic clearance by animals and humans.
- Green toxicology (green chemistry) is aimed at reducing the risk of chemicals by replacing some of the current high-hazard chemicals with new, low-hazard chemicals. Hazards can never be fully eliminated, though; there are always residual risks.
- *Epilogue*: It's not always an advantage to be at the head of the food chain.

21

CONCLUSIONS AND OUTLOOK

Mythbusters

To wrap things up, here are some common prejudices and misconceptions that, hopefully, we have weeded out in the previous chapters of this book. Some are obviously exaggerated to drive the point home. In random order:

> *When we can detect a known toxic chemical in our food or water, soil, or air, it poses a human health risk.* BUSTED.
> *Natural chemicals are less dangerous than synthetic, human-made chemicals.* BUSTED.
> *Herbal dietary supplements and nutraceuticals are safe because they have been tested.* BUSTED.
> *The hazard (potency) of a chemical solely determines the risk.* BUSTED.
> *Studies in lab animals are useless.* BUSTED.
> *All pesticides are highly hazardous.* BUSTED.
> *A good drug should have no adverse reactions.* BUSTED.
> *Exposure limits set by regulatory agencies are valid forever and shouldn't be changed.* BUSTED.
> *Foods without additives are free of toxicants.* BUSTED.
> *If a toxic response, or a disease, strongly correlates with the presence of a specific chemical, it must be caused by that chemical.* BUSTED.
> *All adverse drug reactions can be predicted.* BUSTED.
> *The human health risk from toxic chemicals is beyond our control.* BUSTED.
> *Companies should only be allowed to produce drugs, consumer products, pesticides, or other chemicals that are 100% safe and have zero risk.* NOT POSSIBLE.
> *Wood smoke is cozy.* BUSTED.
> *Metals are harmless; we use them every day.* BUSTED.
> *Reports about chemical risks on social media platforms, the Internet, newspapers, or on TV are reliable and educational for the ordinary consumer because they're communicated in plain language.* NO COMMENT.

Hazard, Exposure, and Risk Resumed

So, what have we learned?

Hopefully, a few key points will stick to the reader's mind who's been so courageous as to read throughout the entire book. The most important point is that exposure and dose are equally crucial for the assessment of a human health risk as the hazard of a chemical. Without a quantitative view, we cannot gauge a chemical risk. This statement may sound trite, but you would be stunned to see how often this simple rule is being ignored. Other variables must be factored in as well, of course, but the dose/exposure is key (see Figure 21.1).

Some of the most tenacious myths addressed in the Introduction have been busted. A particularly prevalent one is the belief that a toxic chemical will harm us just because it can be detected. Relax, turn off the strident alarm, and let's do the math first.

Another ineradicable myth is the conviction that natural chemicals must be less harmful than human-made chemicals. That's not the case; e.g., plants have evolved to contain a large variety of potentially harmful chemicals to ward off predators or protect themselves from insect infestation or fungal infection—they are natural pesticides, which may sound like an oxymoron for people who haven't given much thought about it. We cannot avoid being exposed to naturally occurring chemicals. As to synthetic chemicals, including pesticides and drugs, they are rigorously regulated, and companies must submit a comprehensive safety package before the governmental authorities will consider approval.

At the level of prediction of a human health risk for a specific chemical, we must rely on scientific evidence. This evidence mostly stems from animal data and human epidemiological analysis. For

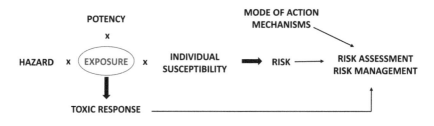

Figure 21.1   The pivotal role of exposure and dose in gauging the human health risk.

short-term (acute) toxicity, it is generally agreed that animal tests are a good predictor for human health risks, after the necessary corrections for species-specific metabolism, dose, exposure route, and other variables, as discussed. In contrast, when it comes to long-term (chronic) toxicity, it is much more difficult to evaluate human health risk. In many cases, animals (rats and mice) are not credible human surrogates because scientific evidence is lacking. Therefore, it is extremely difficult to make regulatory decisions based on, e.g., rodent lifetime bioassays, run with high doses of a chemical, especially when no controlled epidemiological data are available, e.g., from occupational exposure. The near future will likely witness some modifications in the regulatory guidelines for these standardized two-year bioassays that entail great expenditures of laboratory animals, money, and time.

Finally, what experts consider the most challenging chemical threats for humans and our environment are not necessarily the ones that show up in the daily headlines. The most challenging threats are global issues, real problems, and we must tackle them. Without, of course, losing sight of new hitches emerging as new industrial chemicals, pharmaceutical drugs, agrochemicals, and other products hit the market each year.

One thing is clear, though: we can never reduce exposure to chemicals down to zero, and there is no such thing as zero risk. The art of the science of toxicology is to detect the hazards, grasp the dose-response, understand the mechanisms, predict and prevent the toxicity, and manage the risk to human health and our precious environment.

FINAL TAKE-HOME MESSAGE

- Don't panic about bad news from sensational headlines about another slow-killer chemical, or an insidious poison, or a crippling contaminant. Instead, try to get to the bottom of it.
- Whenever possible, read the original reports (reviews or abstracts), or consult the websites of the local authorities (e.g., EPA, FDA).

- Factor in the available data on hazard and potency, and bring information about toxicokinetics into your appraisal.
- Look for *quantitative* data (e.g., TDI, RfD, and RfC) and contrast them with the amounts/concentrations of a specific chemical found in our food, water, air, or environment.
- Estimate your *exposure* and assess your health risk under realistic conditions.
- Balance risk versus benefit (e.g., drugs, foods, and consumer products). There's always a price.
- *Postscript*: Please trust the science.

Appendix 1

Units and Concentrations

Weight: 1 kilogram (kg) = 1000 gram (g); 1 g = 1000 milligram (mg); 1 mg = 1000 microgram (μg); 1 μg = 1000 nanogram (ng); 1 ng = 1000 picogram (pg)

  [1 kg = 35.3 ounces]

Volume: 1 liter (L) = 1000 milliliter (mL); 1 mL = 1000 microliter (μl)

  [1 L = 0.264 gallon]

ppm (parts per million): a measure of a concentration, equal to mg/kg (or mg/L)

ppb (parts per billion): equal to μg/kg (or μg/L)

ppt (parts per trillion): equal to ng/kg (or ng/L)

Appendix 2

Classification of Human Carcinogens as Defined by IARC

The assignment of chemicals to different classes of human carcinogens may be confusing or even misleading for people not familiar with the terminology. For example, when a certain chemical is labeled a *"probable* human carcinogen," this may cause unease or anxiety. It does not mean that you "probably" will get cancer when being exposed to it. It only means that animal data may support the possibility that the chemical may be a human carcinogen, but that human data are not sufficient to make a judgment.

The classification of agents by the International Agency for Research on Cancer (IARC) is a widely accepted standard. It is based on animal studies, human evidence, and mechanistic studies (of how a chemical might act, e.g., as an initiating or promoting agent). There is no quantitative significance; the classes only describe *different levels of evidence.*

Group 1—Carcinogenic to humans. There is sufficient evidence in both humans and animals. *Examples: aflatoxin, arsenic, benzene, benzopyrene, cadmium, TCDD ("dioxin").*

Group 2A—Probably carcinogenic to humans. There is limited evidence in humans and sufficient evidence in animals. *Examples: acrylamide, glyphosate.*

Group 2B—Possibly carcinogenic to humans. There is limited evidence in humans and less than sufficient evidence in animals. *Example: DDT.*

Group 3—Not classifiable as to its carcinogenicity in humans. There is inadequate evidence in humans and inadequate or limited evidence in animals. *Example: atrazine.*

Group 4—Probably not carcinogenic to humans. There is evidence suggesting a lack of carcinogenicity in both humans and animals.

Glossary and Abbreviations

Acceptable daily intake (ADI): the daily intake of a chemical, which during an entire lifetime appears to be without appreciable risk.

Action level: level of a contaminant for enforcement by the food regulatory agencies. It refers to inadvertent residues, e.g., toxins from molds (in contrast to tolerance level).

ADME studies: studies on absorption, distribution, metabolism, and excretion (all parts of pharmacokinetics).

Adverse drug reaction (ADR): any undesirable and harmful effect from a drug (not to be confounded with *side effect* of a drug, which can be beneficial or harmful and which is beyond the intended therapeutic effect).

ALARA: as low as reasonably achievable.

Antioxidant: a foreign or natural agent that counteracts pro-oxidants. It may help reduce or prevent oxidant stress, or even repair oxidative damage to biomolecules.

Bioaccumulation: net accumulation and storage of a chemical in an organism (often in the fat tissue) exceeding the rate of its excretion.

Bioactivation: metabolic conversion of an otherwise harmless chemical to a toxic metabolite.

Biomagnification: increasing levels of a persistent chemical along the food chain.

Body burden: the total amount (or concentration) of a specific chemical in an organism at a given time.

Cancer slope factor (SF): a factor derived from low-dose cancer data in animals to calculate the increased cancer risk from lifetime exposure to a chemical. The greater the slope factor, the higher the potency of the cancer-causing chemical.

Data-poor chemicals: chemicals for which a sound toxicological database is insufficient or absent.

Dietary supplements: pure substances taken in addition to the regular diet to provide additional health benefits. No approval by the health authorities is required, nor a demonstration of efficacy *in vivo*.

Dose scaling: converting a dose that was detected in an animal model to an equivalent dose in humans, based on body surface area.

Endocrine disruptors: chemicals that can interfere with the hormonal (endocrine) systems of the body, with the potential to cause developmental abnormalities.

Exposure: the amount (or concentration) of a chemical around us that might come into contact with an organism.

Forever chemicals: chemicals that are neither degraded in the environment nor significantly metabolized by animals and humans. They persist in the environment for years and decades.

Genetic polymorphism: different variants of the same gene in the population, with possible implications on the toxic response.

Glutathione: a protective biomolecule, present in all cells, that can bind chemically reactive metabolites, thereby inactivating them.

Good Clinical Practice (GCP): international quality standard that delineates the ethical guidelines and provides the technical and scientific framework for clinical trials, making the design, recording, and reporting of the data transparent for the public.

Good Laboratory Practice (GLP): standardized protocols enabling data reproducibility, reliability, data recording and reconstruction of non-clinical studies.

GRAS: Generally Recognized as Safe; a category of food additives that does not need to be tested and for which no tolerance limits are set.

Green toxicology/Green chemistry: the design, production, and use of chemicals that are sustainable and have a lower hazard, and pose smaller human health and environmental risk than traditional chemicals.

Hormesis: a biphasic dose response to chemicals, characterized by a beneficial (or stimulating) effect at low dose and a toxic (or inhibitory) effect at high dose.

IARC: International Agency for Research on Cancer, a section of the WHO of the United Nations.

Idiosyncratic drug reactions: rare, unpredictable adverse reactions to drugs characteristic for an individual. Often (but not always) immune-mediated.

In silico: Computer-simulated studies to predict toxic effects from the structure of a chemical.

In vitro: experiments performed in a Petri dish or cell culture flask (not in live organisms), using cultured cells or isolated organs.

In vivo: studies done in live organisms.

LD$_{50}$: the median dose of a chemical that is lethal for 50% of test animals in a group; an outdated, inaccurate method to estimate the acute toxicity of a chemical.

Legacy chemicals: chemicals that are no longer in active use because of their human health and/or environmental risk.

Margin of exposure (MOE): the ratio of the NOAEL (obtained from animal studies) to the predicted or estimated human exposure level.

Margin of safety (MOS): often used synonymously with MOE, but in drug toxicology often defined as the ratio of the lethal dose in 1% of animals and the effective dose in 99% of the animals (LD_1/LD_{99}).

Maximum contaminant levels (MCL): the highest level of contaminant allowed in the drinking water.

Maximum residue limit (MRL): the highest level of a pesticide that is legally tolerated in or on food when the pesticide is applied correctly.

Maximum tolerated dose (MTD): the highest dose of a chemical administered over the entire lifespan to a laboratory rodent (typically in cancer studies) that does not result in severe toxicity or mortality.

Mechanisms: the detailed cellular and molecular events that underly a specific toxic response.

Metabolism: chemical reactions in the body that are essential for life (e.g., synthesis of biomolecules or energy production). *Metabolism of xenobiotics* means something different: the chemical modification of foreign chemicals. The resulting metabolites may be less toxic or more toxic than the parent compound and are normally more water-soluble and thus excreted more easily.

Microbiome: the total genome of the community of microorganisms normally residing in our body, in particular the intestines. These microorganisms are responsible for breaking down many foods, metabolizing chemicals, and are important for immune defense. The older term "gut flora" has largely been replaced by the newer term, microbiota.

Mode of action: the key events and processes that lead to a toxic response.

Neonicotinoids ("neonics"): a group of synthetic insecticides, structurally similar to nicotine, that are toxic to the nervous system of insects.

No-observed-adverse-effect level (NOAEL): the highest dose level at which a specified toxic response is not observed, but above which toxicity would be seen.

NSAIDs: Non-steroidal anti-inflammatory drugs, a family of pain-killers (e.g., aspirin, diclofenac, and ibuprofen).

Nutraceuticals: products made from food, aimed at preventing or curing a disease. They are not regulated by the health authorities, nor is demonstration of efficacy required.

Off-label use of drugs: use of pharmaceutical drugs for a non-approved condition, or in a non-approved dosage or route of administration.

Omics: a collective term of techniques (with the suffix -omics) that analyze the total of different RNAs or proteins (or other cell products) following exposure to a chemical. For example, the term transcriptomics is the analysis of all newly produced RNAs (transcripts) from activated genes; proteomics is the analysis of all newly produced proteins.

Oxidant stress (aka oxidative stress): an imbalance between pro-oxidants and antioxidants in favor of pro-oxidants, which may lead to oxidative damage of cellular constituents including DNA.

Potency: the relative "power" of a chemical to induce a certain toxic response.

Radicals: chemicals that have gained or ceded an electron and now have an unpaired electron, which makes them incomplete and reactive. The older term "free radicals" is outdated.

Reference dose (RfD): often used synonymously with ADI; a quantitative estimate of daily exposure over lifetime to an agent that is assumed to produce no detectable health effects in humans. The lower the RfD, the higher is the potency of the chemical.

Risk: the *chance* of something bad or harmful happening; risk assessment is the process of *estimating the chance* of some adverse effect to happen.

Systems toxicology: a quantitative integration of toxicological data from *in vivo* and *in vitro* studies including gene expression data (transcriptomics, proteomics).

Threshold of Toxicological Concern (TTC): pragmatic approach to qualitatively assess the risk of low-level substances in diet.

Tolerable Daily Intake (TDI): upper limit for the amount of a chemical, below the level thought to cause adverse health effects.

Tolerance level: level of contaminant on foods as a direct result of proper application of a pesticide (in contrast to action level).

Toxicodynamics: the dynamic interactions of a chemical with a biological target and its downstream biological effects (what a chemical does to the body).

Toxicokinetics: the changes in concentrations of a chemical compound in the organism over time (what the body does to the chemical).

Suggestions for Further Reading

Note: Unlike in the professional scientific literature, where all cited data must be substantiated by providing the original source, the references in this book are selective rather than exhaustive as they seem typical and relevant for the specific context. Also, the individual researchers scattered throughout the text are mentioned in an exemplary way, representing the wealth of other scientific contributions that simply cannot be described within the scope of this book. I hope all those not specifically mentioned feel sympathetic toward the approach chosen.

Chapter 1—Introduction and Chapter 2—What Does "Toxic" Mean?

A useful educational website is the Toxicology Education Foundation (www.toxedfoundation.org).

Boelsterli, U.A. *Mechanistic Toxicology*. Second Ed., CRC Press, Boca Raton, FL, 2007.

Klaassen, C.D. (Ed.) *Casarett and Doull's Toxicology. The Basic Science of Poisons*. Nineth Ed., McGraw Hill, New York, NY, 2018.

Chapter 3—Paracelsus Reloaded: The Dose Concept

Borzelleca, J.F. Paracelsus: Herald of modern medicine. *Toxicol. Sci.* 53: 2–4 (2000). https://doi.org/10.1093/toxsci/53.1.2

Chapter 4—Exposure: The Key Determinant in Risk Assessment

Dekant, W. Metal salts with low oral bioavailability and considerable exposures from ubiquitous background: Inorganic aluminum salts as an example for issues in toxicity testing and data interpretation. *Toxicol. Lett.* 314: 1–9 (2019). https://doi.org/10.1016/j.toxlet.201907.013

Sanajou, S., Sahin, G., and Baydar, T. Aluminium in cosmetics and personal care products. *J. Appl. Toxicol.* 41: 1704–1718 (2021). https://doi.org/10.1002/jat.4228

Scientific Committee on Consumer Safety (EC-SCCS). Opinion on the safety of aluminium in cosmetic products. Submission II. 2020. Available at: https://health.ec.europa.eu/system/files/2021-11/sccs_o_235.

Chapter 5—Natural and Synthetic Chemicals

Breinlinger, S., Phillips, T.J., Haram, B.N., Mares, J., Martinez Yerena, J.A., Hrouzek, P., Sobotka, R., Henderson, W.M., Schmieder, P., Williams, S.M., Lauderdale, J.D., Wilde, H.D., Gerrin, W., Kust, A., Washington, J.W., Wagner, C., Geier, B., Liebeke, M., Enke, H., Niedermeyer, T.H., and Wilde, S.B. Hunting the eagle killer: A cyanobacterial neurotoxin causes vacuolar myelinopathy. *Science* 371: eaax9050 (2021). https://doi.org/10.1126/science.aax9050

Mie, A., Andersen, H.R., Gunnarsson, S., Kahl, J., Kesse-Guyot, E., Rembiałkowska, E., Quaglio, G., and Grandjean. P. Human health implications of organic food and organic agriculture: A comprehensive review. *Environ. Health* 16: 111 (2017). https://doi.org/10.1186/s12940-017-0315-4

Tanner, C.M., Kamel, F., Ross, G.W. *et al.* Rotenone, paraquat, and Parkinson's disease. *Environ. Health Perspect.* 119: 866–872 (2011). https://doi.org/10.1289/ehp.1002839

Chapter 6—What Our Body Does to a Chemical

LoGuidice, A., Wallace, B.D., Bendel, L., Redinbo, M.R., and Boelsterli, U.A. Pharmacologic targeting of bacterial beta-glucuronidase alleviates nonsteroidal anti-inflammatory drug-induced enteropathy in mice. *J. Pharmacol. Exp. Ther.* 341: 447–454 (2012). http://dx.doi.org/10.1124/jpet.111.191122

Wallace, B.D., Wang, H., Lane, K.T., Scott, J.E., Orans, J., Koo, J.S., Venkatesh, M., Jobin, C., Yeh, L.A., Mani, S., and Redinbo, M.R. Alleviating cancer drug toxicity by inhibiting a bacterial enzyme. *Science* 330: 831–835 (2010). https://doi.org/10.1126/science.1191175

Wang, X., Wang, X., Zhu, Y., and Chen, X. ADME/T-based strategies for paraquat detoxification: Transporters and enzymes. *Environ. Pollution* 291: 118137 (2021). https://doi.org/10.1016/j.envpol.2021.118137

Chapter 7—What a Chemical Does to Our Body

Hartwig, A., Arand, M., Epe, B., Guth, S., Jahnke, G., Lampen, A., Martus, H.J., Monien, B., Rietjens, I.M.C.M., Schmitz-Spanke, S., Schriever-Schwemmer, G., Steinberg, P., and Eisenbrand, G. Mode of action-based risk assessment of genotoxic carcinogens. *Arch. Toxicol.* 94: 1787–1877 (2020). https://doi.org/10.1007/s00204-020-02733-2

Pflaum, T., Hausler, T., Baumung, Ackermann, S., Kuballa, T., Rehm, J., and Lachenmeier, D.W. Carcinogenic compounds in alcoholic beverages: An update. *Arch. Toxicol.* 90: 2349–2367 (2016). https://doi.org/10.1007/s00204-016-1770-3

Richardson, J.R., Fitsanakis, V., Westerink, R.H.S., and Kanthasamy, A.G. Neurotoxicity of pesticides. *Acta Neuropathol.* 138: 343–362 (2019). https://doi.org/10.1007/s00401-019-02033-9

Richburg, J.H. and Boekelheide, K. Mono-(2-ethylhexyl) phthalate rapidly alters both Sertoli cell vimentin and germ cell apoptosis in young rat testes. *Toxicol. Appl. Pharmacol.* 137: 42–50 (1996). https://doi.org/10.1006/taap.1996.0055

Chapter 8—Defense Shields

Mattson, M.P. Hormesis defined. *Ageing Res. Rev.* 7: 1–7 (2008). https://doi.org/10.1016/j.arr.2007.08.007

Tribull, T.E., Bruner, R.H., and Bain, L.J. The multidrug resistance-associated protein 1 transports methoxychlor and protects the seminiferous epithelium from injury. *Toxicol. Lett.* 142: 61–70 (2003). https://doi.org/10.1016/S0378-4274(02)00485-X

Chapter 9—Correlation and Causality

Chang, W.H., Herianto, S., Lee, C.C., Hung, H., and Chen, H.L. The effects of phthalate ester exposure on human health: A review. *Sci. Total Environ.* 786: 147371 (2021). https://doi.org/10.1016/j.scitotenv.2021.147371

Feige, J.N., Gelman, L., Rossi, D., Zoete, V., Métivier, R, Tudor, C., Anghel, S.I., Grosdidier, A. Lathion, C., Engelborghs, Y., Michielin, O., Wahli, W., and Desvergne, B. The endocrine disruptor monoethyl-hexyl-phthalate is a selective peroxisome proliferator-activated receptor γ modulator that promotes adipogenesis. *J. Biol. Chem.* 282: 19152–19166 (2007). https://doi.org/10.1074/jbcM702724200

Matthews, R. Storks deliver babies ($p = 0.008$). *Teaching Statist.* 22: 36–38 (2000). https://doi.org/10.1111/1467-9639.00013

Zhu, X., Yin, T., Yue, X., Liao, S., Cheang, I., Zhu, Q., Yao, W., Lu, X., Shi, S., Tang, Y., Zhou, Y., Li, X., and Zhang, H. Association of urinary phthalate metabolites with cardiovascular disease among the general adult population. *Environ. Res.* 202: 111764 (2021). https://doi.org/10.1016/j.envres.2021.111764

Chapter 10—Pesticides: Killers with a License

Agostini, L.P., Dettogni, R.S., dos Reis, R.S., Stur, E., dos Santos, E.V.W., Ventorim, D.P., Garcia, F.M., Cardoso, R.C., Graceli, J.B., and Louro, I.D. Effects of glyphosate exposure on human health: Insights from epidemiological and in vitro studies. *Sci. Total Environ.* 705: 135808 (2020). https://doi.org/10.1016/j.scitotenv.2019135808

Casida, J.E. and Durkin, K.A. Pesticide chemical research in toxicology: Lessons from Nature. *Chem. Res. Toxicol.* 30: 94–104 (2017). https://doi.org/10.1021/acs.chemrestox.6b00303

Kassotis, C.D., Vandenberg, I.N., Demeneix, B.A., Porta, M., Slama, R., and Trasande, L. Endocrine-disrupting chemicals: Economic, regulatory, and policy implications. *Lancet Diabetes Endocrinol.* 8: 719–730 (2020). https://doi.org/10.1016/S2213-8587(20)30128-5

Meftaul, I.M., Venkateswarlu, K., Dharmarajan, R., Annamalai, P., Asaduzzaman, M., Parven, A., and Megharaj, M. Controversies over human health and ecological impact of glyphosates: Is it to be banned in modern agriculture? *Environ. Pollut.* 263: 114372 (2020). https://doi.org/10.1016/j.envpol.2020.114372

Oerke, C.E. Crop losses to pests. *J. Agricult. Sci.* 144: 31–43 (2006). https://doi.org/10.1017/S0021859605005708

Weidenmüller, A., Meltzer, A., Neupert, S., Schwarz, A., and Kleineidam, C. Glyphosate impairs collective thermoregulation in bumblebees. *Science* 376: 1122–1126, (2022). https://doi.org/10.1126/science.abf7482

Zoller, O., Rhyn, P., Rupp, H., Zarn, J.A., and Geiser, C. Glyphosate residues in Swiss market foods: Monitoring and risk evaluation. *Food Addit. Contam. B* 11: 83–91 (2018). https://doi.org/10.1080/19393210.2017.1419509

Chapter 11—Toxic Food

Dolan, L.C., Matulka, R.A., and Burdock, G.A. Naturally occurring food toxins. *Toxins* 2: 2289–2332 (2010). https://doi.org/103390/toxins2092289

Eisenbrand, G. Revisiting the evidence for genotoxicity of acrylamide (AA), key to risk assessment of dietary AA exposure. *Arch. Toxicol.* 94: 2939–2950 (2020). https://doi.org/10.1007/s00204-020-02794-3

Schrenk, D., Fahrer, J., Allemang, A., Fu, P., Lin, G., Mahony, C., Mulder, P.P.J., Peijnenburg, A., Pfuhler, S., Rietjens, I.M., Sachse, B., Steinhoff, B., These, A., Troutman, J, and Wiesner, J. Novel insights into pyrrolizidine alkaloid toxicity and implications for risk assessment: Occurrence, genotoxicity, toxicokinetics, risk assessment—A workshop report. *Planta Med.* doi 10.1055/a-1646-3618 (2021). https://doi.org/10.1055/a-1646-3618

Tareke, E., Rydberg, P., Karlsson, P., Eriksson, S., and Törnqvist, M. Analysis of acrylamide, a carcinogen formed in heated foodstuffs. *J. Agric. Food Chem.* 50: 4998–5006 (2002). https://doi.org/10.1021/jf020302f

Timmermann, C.A.G., Molck, S.S., Kadawathagedara, M., Bjerregaard, A.A., Törnqvist, M., Brantsaeter, A.L., and Pedersen, M. A review of dietary intake of acrylamide in humans. *Toxics* 9, 155 (2021). https://doi.org/10.3390/toxics9070155

Chapter 12—Dietary Supplements: The More the Better?

Halliwell, B. Are polyphenols antioxidants or pro-oxidants? What do we learn from cell culture and *in vivo* studies? *Arch. Biochem. Biophys.* 476: 107–112 (2008). https://doi.org/10.1016/j.abb.2008.01.028

Hudson, A., Lopez, E., Almalki, A.J., Roe, A.L., and Calderon, A.I. A review of the toxicity of compounds found in herbal dietary supplements. *Planta Med.* 84: 613–626 (2018). https://doi.org/10.1055/a-0605-3786

Ronis, M.J., Pedersen, K.B., and Watt, J. Adverse effects of nutraceuticals and dietary supplements. *Annu Rev. Pharmacol. Toxicol.* 58: 583–601 (2018). https://doi.org/10.1146/annurev-pharmtox-010617-052844

Shaito, A., Posadino, A.M., Younes, N., Hasan, H., Halabi, S., Alhababi, D., Al-Mohannadi, A., Abdel-Rahman, W.M., Eid, A.H., Nasrallah, G.K., and Pintus, G. Potential adverse effects of resveratrol: A literature review. *Int. J. Mol. Sci.* 21: 2084 (2020). https://doi.org/10.3390/ijms21062084

Chapter 13—Significant Chemical Risks: Persistent and Widespread

Cohen, S.M., Arnold, L.L., and Tsuji, J.S. Inorganic arsenic: A nongenotoxic threshold carcinogen. *Curr. Opin. Toxicol.* 14: 8–13 (2019). https://doi.org/10.1016/j.cotox2019.05.002

Fullerton, D.G., Bruce, N., and Gordon, S.B. Indoor air pollution from biomass fuel smoke is a major health concern in the developing world. *Transact. Royal Soc. Trop. Med. Hyg.* 102: 843–851 (2008). https://doi.org/10.1016/j.trstmh.2008.05.028

Lamm, S.H., Robbins, S., Chen, R., Lu, J., Goodrich, B., and Feinleib, M. Discontinuity in the cancer slope factor as it passes from high to low exposure levels—arsenic in the BFD-endemic area. *Toxicology* 326: 25–35 (2014). http://dx.doi.org/10.1016/j.tox.2014.08.014

Naeher, L.P., Brauer, M., Lipsett, M., Zelikoff, J.T., Simpson, C.D., Koenig, J.Q., and Smith, K.R. Woodsmoke health effects: A review. *Inhal. Toxicol.* 19: 67–106 (2007). https://doi.org/10.1080/08958370600985875

Yang, L., Zhang, Y., Wang, F., Luo, Z., Guo, S. Toxicity of mercury: Molecular evidence. *Chemosphere* 245: 125586 (2020). https://doi.org/10.1016/j.chemosphere.2019125586

Chapter 14—Drugs

Craveiro, N.S., Lopes, B.S., Tomas, L., and Almeida, S.F. Drug withdrawal due to safety: A review of the data supporting withdrawal decision. *Curr. Drug Safety* 15: 4–12 (2020). https://doi.org/10.2174/1574886314666191004092520

Iasella, C.J., Johnson, H.J., and Dunn, M.A. Adverse drug reactions: Type A (intrinsic) or Type B (idiosyncratic). *Clinics Liver Dis.* 21: 73–87 (2017). http://dx.doi.org/10.1016/j.cld.2016.08.005

Volkow, N.D. and Blanco, C. The changing opioid crisis: Development, challenges and opportunities. *Mol. Psychiatry* 26: 218–233 (2021). https://doi.org/10.1038/s41380-020-0661-4

Chapter 15—Safety Assessment

Benigni, R. Bassan, A., and Pavan, M. *In silico* models for genotoxicity and drug regulation. *Expert Opin. Drug Metab. Toxicol.* 16: 651–662 (2020). https://doi.org/10.1080/17425255.20201785428

Ching, T., Toh, Y.C., Hashimoto, M., and Zhang, Y.S. Bridging the academia-to-industry gap: Organ-on-a-chip platforms for safety and toxicology assessment. *Trends Pharmacol. Sci.* 42: 715 (2021). https://doi.org/10.1016/j.tips.2021.05.007

Maharao, N., Antontsev, V., Wright, M., and Varshney, J. Entering the era of computationally driven drug development. *Drug Metab. Rev.* 52: 283–298 (2020). https://doi.org/10.1080/03602532.2020.1726944

Nair, A.B. and Jacob, S. A simple practice guide for dose conversion between animals and human. *J. Basic Clin. Pharm.* 7: 27–31 (2016). https://doi.org/10.4103/0976-0105.177703

Reagan-Shaw, S., Nihal, M., and Ahmad, N. Dose translation from animal to human studies revisited. *FASEB J.* 22: 659–661 (2007). https://doi.org/10.1096/fj.07-9574LSF

Watkins, P.B. DILIsym: Quantitative systems toxicology impacting drug development. *Curr. Opin. Toxicol.* 23: 67–73 (2020). https://doi.org/10.1016/j.cotox.2020.06.003

Chapter 16—Acceptable Limits, Tolerance, and Red Lines

More, S.J., Bampidis, V., Benford, D., Bragard, C., Halldorsson, T.I., Hernandez-Jerz, A.F., Bennekou, S.H., Koutsoumanis, K.P., Machera, K., Naegeli, H., Nielsen, S.S., Schlatter, J.R., Schrenk, D., Silano, V., Turck, D., Younes, M., Gundert-Remy, U., Kass, G.E.N., Kleiner, J., Rossi, A.M., Serafimova, R., Reilly, L., and Wallace, H.M. Guidance on the use of the Threshold of Toxicological Concern approach in food safety assessment. *EFSA J.* 17: 5708 (2019). https://doi.org/10.2903/j.efsa.2019.5708

Chapter 17—Risk Assessment

Ames, B.N., Profet, M., and Gold, L.S. Dietary pesticides (99.99% all natural). *Proc. Natl. Acad. Sci. USA* 87: 7777–7781 (1990). https://doi.org/10.1073/pnas87.19.7777

Cichocki, J.A., Smith, G.J., Mendoza, R., Buckpitt, A.R., Van Winkle, L.S., and Morris, J.B. Sex differences in the acute nasal antioxidant/antielectrophilic response of the rat to inhaled naphthalene. *Toxicol. Sci.* 139: 234–244 (2014). https://doi.org/10.1093/toxsci/kfu031

Lee, J.E.C., Lemyre, L., Mercier, P., Bouchard, L., and Krewski, D. Beyond the hazard: The role of beliefs in health risk perception. *Hum. Ecol. Risk. Assess.* 11: 1111–1126 (2005). https://doi.org/10.1080/10807030500278636

Smith, C.J. and Perfetti, T.A. The "false-positive" conundrum in the NTP 2-year rodent cancer study database. *Toxicol. Res. Appl.* 2: 1–13 (2018). https://doi.org/10.1177/2397847318772839

Wang, N.C.Y., Venkatapathy, R., Bruce, R.M., and Moudgal, C. Development of quantitative structure-activity relationship (QSAR) models to predict the carcinogenic potency of chemicals. II. Using oral slope factor as a measure of carcinogenic potency. *Reg. Toxicol. Pharmacol.* 59: 215–226 (2011). https://doi.org/10.1016/j.yrtph.2010.09.019

Yost, E.E., Galizia, A., Kapraun, D.F., Persad, A.S., Vulimiri, S.V., Angrish, M., Lee, J.S., and Druwe, I.L. Health effects of naphthalene exposure: A systematic evidence map and analysis of potential considerations for dose-response evaluation. *Envir. Health Perspect.* 129: 076002 (2021). https://doi.org/10.1289/EHP7381

Chapter 18—Gauging the Risk Against the Benefit

Huang, Y., Law, J.C., Lam, T.K., Leung, K.S. Risks of organic UV filters: A review of environmental and human health concern studies. *Sci. Total Environ.* 755: 142486 (2021). https://doi.org/10.1016/j.scitotenv.2020.142486

Kinnberg, K.L., Petersen, G.I., Albrektsen, M., Minghlani, M., Awad, S.M., Holbech, B.F., Green, J.W., Bjerregaard, P., and Holbech, H. Endocrine-disrupting effects of the ultraviolet filter benzophenone-3 in zebrafish, *Danio rerio. Environ. Toxicol. Chem.* 34: 2833–2840 (2015). https://doi.org/10.1002/etc.3129

Nash, J.F. Human safety and efficacy of ultraviolet filters and sunscreen products. *Dermatol. Clin.* 24: 35–51 (2006). https://doi.org/10.1016/j.det.2005.09.006

Rembold, C.M. The health benefits of eating salmon. *Science* 305: 475 (2004).

Reiss, R., Johnston, J., Tucker, K., DeSesso, J.M., and Keen, C.L. Estimation of cancer risks and benefits associated with a potential increased consumption of fruits and vegetables. *Food Chem. Toxicol.* 50: 4421–4427 (2012). https://doi.org/10.1016/j.fct.2012.08.055

Valcke, M., Bourgault, M.H., Rochette, L., Normandin, L., Samuel, O., Belleville, D., Blanchet, C., and Phaneuf, D. Human health risk assessment on the consumption of fruits and vegetables containing residual pesticides: A cancer and non-cancer risk/benefit perspective. *Environ. Intern.* 108: 63–74 (2017). https://doi.org/10.1016/j.envint.2017.07.023

Chapter 19—Risk Communication

Solomon, K.R., Velders, G.J.M., Wilson, S.R., Madronich, S., Longstreth, J., Aucamp, P.J., and Bornman, J.F. Sources, fates, toxicity, and risks of trifluoroacetic acid and its salts: Relevance to substances regulated under the Montreal and Kyoto Protocols. *J. Toxicol. Environ. Health, Part B.* 19: 289–304 (2016). https://doi.org/10.1080/10937404.2016.1175981

Chapter 20—Toxicological Challenges

Aschner, M., Autrup, H.N., Berry, Sir C.L., Boobis, A.R., Cohen, S.M., Creppy, E.E., Dekant, W., Doull, J., Gallli, C.L., Goodman, J.I., Gori, G.B., Greim, H.A., Joudrier, P., Kaminski, N.E., Klaassen, C.D., Klaunig, J.E., Lotti, M., Marquart, H.W.J., Pelkonen, O., Sipes, I.G., Wallace, K.B., and Yamazaki, H. Upholding science in health, safety and environmental risk assessments and regulations. *Toxicology* 371: 12–16 (2016). https://doi.org/10.1016/j.tox.2016.09.005

Colnot, C. and Dekant, W. Issues in the hazard and risk assessment of perfluoroalkyl substance mixtures. *Toxicol. Lett.* 353: 79–82 (2021). https://doi.org/10.1016/j.toxlet.2021.10.005

Danopoulos, E., Jenner, L.C., Twiddy, M., and Rotchell, J.M. Microplastic contamination of seafood intended for human consumption: A systematic review and meta-analysis. *Environ. Health Perspect.* 128: 126002 (2020). https://doi.org/10.1289/EHP7171

Leslie, H.A., van Velzen, M.J.M., Brandsma, S.H., Vethaak, A.D., Garcia-Vallejo, J.J., and Lamoree, M.H. Discovery and quantification of plastic particle pollution in human blood. *Environ. Intern.* 163: 107199 (2022). https://doi.org/10.1016/j.envint.2022.107199

Lim, X.Z. Microplastics are everywhere—but are they harmful? *Nature* 593: 22–25 (2021). https://doi.org/10.1038/d41586-021-01143-3

Panieri, E., Baralic, K., Djukic-Cosic, D., Djordevic, A.B., and Saso, L. PFAS molecules: A major concern for the human health and the environment. *Toxics* 10: 44 (2022). https://doi.org/10.3390/toxics10020044

Sunderland, E.M., Hu, X, C., Dassuncao, C., Tokranov, A.K., Wagner, C.C., and Allen, J.G. A review of the pathways of human exposure to poly- and perfluoroalkyl substances (PFASs) and present understanding of health effects. *J. Exp. Sci. Environ. Epidemiol.* 29: 131–147 (2019). https://doi.org/10.1038/s41370-018-0094-1

Yang, W., Wang, L., Mettenbrink, E.M., DeAngelis, P.L., and Wilhelm, S. Nanoparticle toxicology. *Annu. Rev. Pharmacol. Toxicol.* 61: 269–289 (2021). https://doi.org/10.1146/annurev-pharmtox-032320-110338

Acknowledgments

I am extremely grateful to numerous friends, colleagues, and experts, who took the time to vet individual chapters and make me aware of errors or how to improve the overall message. Among these, my sincere thanks go to Patrick Bouis, Marsha L. Butler, Samuel M. Cohen, Steven D. Cohen, Lucio G. Costa, Wolfgang Dekant, Béatrice Desvergne, Daniel Dietrich, Beat Glogger, Bruno Hagenbuch, Werner K. Lutz, Thomas Petry, Stephen M. Roberts, Robert A. Roth, Beat Schmid, Susanne and Christian Sengstag, and Kendall B. Wallace. If there are still mistakes, they are entirely mine.

I also want to thank Stephen M. Zollo and the editorial team of CRC Press for their help and excellent guidance throughout the publication process.

Index

Note: Page references in *italics* denote figure and in **bold** tables.